What is Math?

by

Field Cady

You rock Camille! Thanks for being such an awesome friend and fellow nerd!

Field

Copyright © 2015 Field Cady

All rights reserved.

DEDICATION

To Ryna. Thank you, honey, for always believing in me.

ACKNOWLEDGEMENTS

In writing this book I have often felt like the inspiration came from my wife Ryna, the editing came from a range of friends and family, and the content itself came from the many researchers whose work I have drawn on. I'm sure I contributed something here and there, but I feel like I've mostly been a conduit for the brilliant ideas and hard work of others. To all those who have made this book possible, I am deeply grateful.

CONTENTS

	Acknowledgments	i
1	Dawn on the Savannah	Pg 1
2	What is Language?	Pg 28
3	The Rise of Numbers	Pg 69
4	Proof and Prejudice	Pg 87
5	Math and Science	Pg 112
6	The True, the False, and the Patently absurd	Pg 130
7	Conversations with Silicon	Pg 157
8	Why Nobody Actually Understand Economics	Pg 176
9	The New Age of Empirical Math	Pg 196
	Notes on Figures, Recommended Reading	Pg 222

1 DAWN ON THE SAVANNAH

Mathematics is a language. Obviously it's an unusual language, but there's nothing that makes it fundamentally different from English or Swahili. Maybe math uses some parts of the brain more and others less, and of course we tend to use it for very different things. But at the end of the day these are all just differences of degree. Math is a language. Right?

That's the conclusion I eventually came to when the question "what is math?" was brought up at a party. I was in computer science grad school, and my roommates and I were having a group of friends from the department over to our place to celebrate the holidays. It was a pretty nerdy crowd, but otherwise there was a lot of variety. The largest group of people were the "normal" computer scientists; engineers and problem solvers who really enjoyed making cool things and understanding how they work. There were also the "theory" people, who spent all their time in front of a chalkboard arguing about technical minutia of some new algorithm. The "human computer interaction" crowd did usability studies and pondered new ways for people to interact with technology. They spent a lot of time putting together prototypes, and certainly had the technical chops of a computer scientist, but at heart I always felt that they were more like psychologists or product designers.

Then of course there was me, who had stumbled into computer science almost by accident and knew less about the subject than anybody else in the room. I studied physics in college, but quickly discovered that it was a public safety hazard to let me into a laboratory with lasers and toxic chemicals. So in grad school I gravitated toward mathematics, where the most expensive equipment I was likely to break was a pencil. On a lark I took some computer science seminars, and made the surprising discovery that the math in computer science departments is actually a lot cooler than the math in math departments (more on that later). So I applied to a PhD program at Carnegie Mellon, deftly glossing over the fact that I didn't actually know computer science. It worked, and so here I was.

The well-spiked egg nog had started to take effect at the party, and the conversation drifted in some odd directions. I don't remember who first posed the philosophical question "what is math?", but I do remember my shock that nobody in the room had a coherent answer. The engineers just used math as a way to calculate whatever number or formula they needed. The human-computer-interaction people weren't any better, even though they used all kinds of statistics in running their trials and were (how do I put this tactfully?) a little more socially normal. Even the theorists weren't any help, despite the fact that they lived and breathed math.
People talked vaguely about math being a game of symbols, or a set of rules you follow, but they quickly trailed off into unintelligibility. Bottom line: nobody had an answer.

This book is my answer to the question, which I wish I'd been knowledgeable enough (and, um, sober enough) to come up with during that party. Instead I mulled it over for weeks and months afterward. I eventually concluded that math is a branch of language, not fundamentally different from the words I'm writing right now. More surprisingly though, I realized that this observation isn't just an academic point! It's the key to understanding how humans learn math, how we use it on a daily basis, and how we abuse it - even in situations as dire as economic policy and medical decisions. Seeing that math is a language illuminates every part of the subject, brings it down to earth in a more user-friendly format, and frames it in the bigger context of the human story.

Border Cave is a large rock overhang in southern Africa, perched high in the Lebombo mountains of Swaziland and overlooking the Ingavuma river. It was discovered in 1933 by a man named W.E. Horton, who was exploring the mountains looking for bat guano.

Horton dug out several feet of dirt on the cave floor and sifted through it, looking for the precious droppings which he could sell as fertilizer. Among the usual rocks and sticks he uncovered several large mammal bones, which found their way into the hands of Professor R.A. Dart, an anatomist at the local university. Some of them were from wild animals, as you might expect from an African cave. Others, however, were the remains of stone age humans.

It turns out that humans have lived in Border Cave on-and-off for the last 200,000 years - so far back they weren't even "humans" in the modern sense. The cave is filled with artifacts, left there by different people at different times throughout all those years. They give us snapshots of what life was like during human prehistory, telling the story of how we gradually evolved from bipedal apes, through crude hunter-gatherers, and finally to modern homo sapiens. Border Cave was also the home of the first mathematician in recorded history. Based on what we can tell from the archaeological record, she was a cave woman who lived about 37,000 years ago. Let's call her Hera.

Now 37,000 years is obviously quite a while, so of course we don't have any biographical details about Hera. She lived long before the invention of writing, and even if she found her way into oral tradition it would have faded eons ago. Everything we know about her comes from the detritus in Border Cave. Fortunately though, it tells us a lot. We know that Hera was physically modern, probably with dark skin and black hair. If you put her in modern clothes she would look perfectly normal walking down the street today. She wore clothing made out of furs, and maybe even spun fibers, since the cave has bone needles from her time. She probably adorned herself with beaded shell necklaces, with gems of quartz and chalcedony, and used red ochre clay as makeup.

Border Cave would have made a wonderful home for Hera and her tribe. In the first place it was a cave, which meant permanent, high-quality shelter - not so easy to come by in those days. From the mouth of the cave there was a gorgeous view, looking down the lush slopes of the mountains all the way to the river below.

There were bushpigs, zebras and buffalo to hunt, and roots and berries to gather. The world was going through an ice age at the time, so temperatures would have been 14 degrees fahrenheit cooler than they are now. It would have been a little on the chilly side in that part of Africa, but generally well above freezing - nothing a nice warm buffalo hide couldn't take care of. There would have been at most a few dozen people in Hera's tribe, so there was plenty of food to go around, and life was very comfortable.

Hera's people didn't have any notion of this way of life changing. Since there was no writing system they wouldn't have had any history outside of living memory, except for some semi-mythical oral traditions. Technological progress was creeping along at a snail's pace, so nobody would have noticed it as a trend. In the modern world we're very taken with the "forward march of history", where society is constantly moving on to greater heights and exploring new frontiers. For Hera's people though, this linear progression was replaced by the constant cycling of the days, the seasons and the generations. Her ancestors (or maybe the people who her ancestors had replaced) had dwelt in the same cave, hunting the same animals and living the same life for millennia. Always had been, and always would be. Yet in the midst of this stasis, Hera was doing something revolutionary: she was counting.

That's not to say she could count out loud. Maybe she could, and maybe she couldn't. There's no way to tell from the archaeological record whether her people had words for numbers. But even if they did have a number system it probably wasn't very sophisticated. We know this by observing modern hunter-gatherer societies. Some of them have crude counting systems, usually based on counting fingers and toes and maxing out around twenty. Others are less ambitious, with numbers only up to five or so. Many of them simply have no number words at all.

Despite these likely limitations, Hera understood that numbers can come in any size, and she wanted to keep track of large ones even if her language didn't have words for them. So she used a simple, written notation for numbers that we now call "tallying": she made notches in a baboon leg bone.

When W.E Horton discovered human remains among the bat guano of the cave excavations started in earnest. Ultimately the cave yielded a whopping 69,000 artifacts, a gold mine for the study of early human evolution. Stone tools, jewelry, bones from humans and hunted animals - it was all there, the only things left of Hera and her ancient people. Among the artifacts was the bone that Hera had used for tallying – the only record we have of Hera. It was christened the "Lebombo bone", and has been carbon-dated to 35,000 B.C.E. The Lebombo bone has 29 notches; Hera was counting the days of her period.

I am, of course, taking some major liberties in describing Hera. We know a lot about how her people lived, but as far as Hera herself goes all we have is the notched bone, and *somebody* must have carved it. Hera might actually have been counting the days between full moons rather than her period; in that case she could equally well have been a man. We'll never really know. Chances are that Hera didn't even invent tallying herself; it was probably a standard technique in her tribe, and it's just random chance that hers is the bone that lasted until the present day. The details will always be sketchy, but in any case the Lebombo bone is the oldest unequivocal example we have of counting, which means that mathematics is at least 37,000 years old.

Tallying might seem crude by today's standards, but in the bigger picture it was revolutionary. No other animal in the world – not even the smart, tool-using ones like apes – is known to do anything similar. And humans don't just squeak by; tallying comes so naturally to our brains that it barely even needs to be taught. Of course, learning the modern decimal system or how to do long division still requires work, but it's all built on the same cognitive ability, and this "counting instinct" is almost second nature to humans. Once a species has developed the gray matter to be able to understand tallying, rocket science is just a small step further.

There are several things about tallying that set it apart. The most obvious one is that tallying is symbolic. There is no natural relationship between notches in a bone and days in a month. They don't look or feel or smell like each other; all they have in common is quantity. Hera invented a correspondence between these basically unrelated things, because notches are a lot easier to keep track of. Symbolic abstraction is not quite unique to humans. Animal calls are "symbols" of a sort, and in some animals they are even learned rather than being instinctive. But as impressive as the symbol usage is in some animals, it falls pitifully short of what we take for granted in humans.

The second thing about tallying is that it's "generative". You could have a different word or symbol for each number. A tick mark for one, a cross for two, a circle for three, and so on. Different numbers are different concepts after all, so it makes sense to give each one its own symbol. The problem with such a system, of course, is that you can only count as high as the number of symbols you have. In contrast, a generative system starts with a finite collection of symbols, like the digits 0 through 9 or the tick mark of tallying. It then combines these symbols, according to some set of logical rules, to express *any* number. Four tick marks in a row isn't just repetitions of the number one; it is a whole new quantity, in the same way that 15 is a distinct number from 1 and 5.

This might seem like elementary stuff to you, but that's only because you are part of the most cognitively advanced species the world has ever known. No other creature on Earth understands generative symbol systems. No other creature ever *could* understand them.

The final thing that made Hera's tallying so special is that she was keeping track of something that was difficult to wrap her head around. Humans have a pretty good gut instinct for small numbers. Even newborns can automatically distinguish between one, two and three objects - it comes built-in to our brains. But as the numbers get larger, our intuition gets fuzzier. If somebody showed you a pile of 15 apples and didn't give you time to count them, would you recognize how many there were? I sure wouldn't. I could tell you it was more than 5 and less than 100, but I couldn't tell the difference between 15 and 16 apples without counting them. In the same way, twenty-nine days doesn't "feel" any different from twenty-eight days. Hera wanted to keep track of fine distinctions like this, so she had to invent something that was more precise than the raw human brain.

This might be a good time to introduce what will become a key theme in this book: intuitive scope. This is a term that I use to describe anything that our brains can fully understand without resorting to external tools, like language or counting. Estimating how far you have to reach to grab an apple from the table, reading social dynamics, and recognizing how many fingers a person is holding up fit within intuitive scope. Finding the distance from Manhattan to Argentina, understanding the intricacies of the federal bureaucracy, and counting the people in a crowded room do not. If Hera had never aspired to count higher than four she would never have needed a tally stick. As it is though she was stepping outside of her intuitive scope, and needed some help with that. Intuitive scope is obviously a fuzzy category, but it serves as a useful tool for understanding the themes of this book. We'll see that natural language usually (but not always) stays within the confines of intuitive scope, but mathematics regularly leaves our intuitions in the dust.

So where did Hera get the cognitive chops to step outside of her intuitive scope and invent tallying? It's hard to imagine evolution creating such a remarkable, sophisticated skill just so that our ancestors could count the days in a month. Biological cave lion repellant would probably have been a lot more useful. However, counting isn't the only ability that's unique to humans. Hera and her people had one other major skill going for them, another symbolic skill in fact: they could talk. Unlike counting, the evolutionary benefits of language are obvious: coordinating hunts, navigating social dynamics, passing down knowledge, and so on.
  Is it possible that these two very different activities are actually the same thing under the hood? That's the thesis I'll argue in this

book. At its core mathematics is just a variety (albeit an unusual one) of the single quintessentially human activity: language.

Hera was part of a much larger revolution that was picking up steam at that time, a dramatic shift to what's called behavioral modernity. From the archaeological record we can see that anatomically modern humans lived in Africa as far back as 200,000 years ago, around the time Border Cave was first settled. Skeletal structure aside though, they lived basically the same way their predecessors had - with crude stone tools and fire, but not much else. There is almost nothing in the way of art, culture (outside of making tools) or the other things we associate with our species. A modern skeleton does not a modern human make. But starting around 50,000 B.C.E the archaeological record comes to life. There are figurines, lovingly crafted in the likeness of ancient people, or maybe their gods. Jewelry appears, often made out of goods from far away places. For example, somebody went to a lot of trouble to get seashells from the coast to Border Cave, where they were strung into necklaces. Caves had been homes for a long time, but now they became art galleries as well, blossoming with elaborate murals. Something monumental happened roughly 50,000 years ago, but scientists aren't sure what it was. One possible explanation is the Lake Toba supervolcanoe, which erupted 70,000 years ago and may have reduced the human species down to a few thousand individuals. Evolution happens faster in small populations, so Toba might have opened the door for major changes to take root in some isolated populations. This is speculation though; we don't know exactly what happened. But before that time our ancestors had been unusually intelligent, tool-using apes. This transition - which is sometimes called the Great Leap Forward (GLF) - is what made us fully human.

Right about the time that Hera was notching bones in the southern tip of Africa, somebody a continent away in the south of France was drawing the oldest known (as of this writing) cave paintings, in Chauvet Cave. These paintings weren't just stone-age graffiti. Check it out:

What is Math?

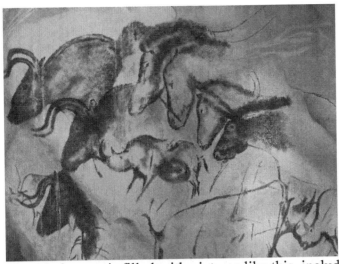

The whole cave is filled with pictures like this, including cattle, reindeer, cave lions, and even rhinoceros (no humans though, interestingly). The details are intricate, the proportions are right, and the overall picture is moving. They even use a special etching technique to create a 3D effect! Looking at the walls of the cave, it's obvious that whoever painted them was a lot more than just mentally modern. He or she (it could also have been a number of individuals, spread over several generations) was a talented and sensitive artist by the standards of any age. Pablo Picasso is reported to have visited a similar cave in Altamira, Spain. He was blown away by the sophistication of the ancient painters and declared that "after Altamira, all is decadence". In 35 millennia, humanity hadn't learned anything about art that wasn't already known at Altamira. Whatever it is that makes our minds so special, the cave artists definitely had it too.

Around the same time, musical instruments start showing up in the archaeological record, in the form of flutes made from animal bones. Unfortunately studying bone flutes is a little tricky, since it can be hard to tell whether something is a full-fledged musical instrument or just an old bone with holes in it; twenty thousand years of weathering has a way of blurring the line. New discoveries are being made constantly, and these words could easily be out of date by the time you read them, but as of this writing, the oldest probable musical instrument is the Divje Babe

flute. It's a broken, 43 thousand year old bear's bone, and its two holes are spaced roughly as far apart as human fingers.
 Researchers believe that the Divje Babe flute was originally part of a larger flute with additional holes. So it looks like music, along with cave painting and tallying, emerged during the Great Leap Forward.

Human beings weren't alone in the world at this time. Our relatives the Neanderthals had been roaming around Europe for the last 300,000 years. DNA evidence suggests that we and Neanderthals are both descended from an earlier species called Homo Heidelbergensis. Heidelbergensis were taller and more muscular than modern humans, with a low-set forehead. Their brains were almost as big as ours, and they were avid tool users, with stone tools and even fire. Heidelbergensis lived in Africa, and about half a million years ago a group of them migrated into Europe. Those who went to Europe evolved into the Neanderthals, while those who stayed in Africa gave rise to modern humans.

Around the time of the GLF, a group of modern humans called the Cro-Magnon was migrating out of Africa and coming into contact with their long-lost cousins the Neanderthals. It doesn't appear to have been a pleasant reunion. Neanderthals were tough and hardy, but they consistently retreated in the face of human advancement.
 By 25,000 years ago the Neanderthals disappeared completely from the archaeological record (although there was a little bit of inter-breeding - modern Europeans are 1-4% Neanderthal). It seems that, through some combination of better survival skills and probably war, we wiped them out.

Why is it that modern humans managed to outcompete the Neanderthals, and now we're the leading species in the world rather than them? It certainly isn't that we were stronger - Neanderthals made Arnold Schwarzenegger look like yours truly.
 Climate change probably had something to do with it, since Neanderthals' diets and hunting practices were adapted to an ice age Europe that was slowly thawing out. But there's a good

chance that's not the whole story, and we should look for other explanations as well.

A lot of stuff happened about 40 thousand years ago. Hera invented tallying, as a tool to help her brain deal with large numbers. At about the same time somebody else was creating the first cave murals in Altamira. Another person was carving the first flutes. Stone figurines made their debut in history. As mathematics and art were blossoming, modern humans were displacing the Neanderthals. Each of these different developments is astonishing in its own right, so it seems a bit much to believe they happened at the same time by coincidence. Is it possible that a single factor caused all of them? Let's examine the possibilities.

One common thread is tool use. Paintbrushes and musical instruments were really just new types of tools. Plus, the Cro-Magnon made better spears and knives than the Neanderthals, which would have helped their hunting as well as giving them an edge in any inter-species wars. Is it possible that the GLF was about tool use?

The problem with this idea is that tools have been around for a long time. As far back as Homo Heidelbergensis, our branch of the tree of life has been making stone weapons and even using fire. The archaeological records shows us that, while the Cro-Magnon were excellent tool users, there was no "great leap forward" in the quality of their knives and spear points. They got better over time, but it was the same gradual improvement that had been going on for millennia. It's true that we invented new tools during the GLF, like paintbrushes and tally sticks, but in those cases the tools themselves were kind of incidental. They were invented as a way to solve new problems, and the real question is why those problems came up in the first place. Ok, so tool use is out.

Let's look at cave paintings and music. Both of them are art forms, but the similarity ends there. They appeal to totally different senses. They are produced in completely different ways.

Music lasts only for the duration of a song, whereas cave paintings are meant to endure forever. Yet for some reason, music and painting seem to have emerged in lock-step. What gives?

Now, I'm not an artist, and by speculating about art I'm venturing outside my realm of competence. But it strikes me that the big thing painting and music have in common is complex communication. They both convey a message, like a story or an emotion. I don't just mean simple, evolution-based messages like "I'm higher in the pecking order than you". They have nuance, details, and layers of meaning. Of course the core message might still be something simple, like "I'm a good mate" (think modern pop), but even then it's fleshed out in extraordinary detail and told in a novel way.

When you see it this way the GLF starts to look like a revolution not in tools, but in the communication of ideas. It suddenly became important to people to express their emotions through music, or to immortalize their stories on the walls of caves. Art like this isn't just aesthetically pleasing. It's expressive. Tally sticks aren't really art, but they fall under the communication umbrella too. In all of these cases the artist or tallier has something to say, in a way that it has never been said before. And of course, saying things is what homo sapiens does best. We aren't the only creatures that use tools, that walk upright or that have opposable thumbs. We are, however, the only species with anything resembling a modern language. There's no analogue anywhere else in the animal kingdom to the systematic, limitless ways that humans can construct a sentence to get their point across. There is, however, a parallel in how we combine musical notes into ballads, or brush strokes into murals. Could it be that language, and the cognitive leaps that it required, is the switch that got flipped in the Great Leap Forward?

I would love to answer this question definitively, but sadly there's no realistic chance of it. Language doesn't leave any physical evidence, so all we have is educated guesswork.

However, language does tie all the various parts of the GLF together into one coherent narrative. If the GLF truly marked the beginning of language, then it makes perfect sense that our ancestors suddenly wanted to talk through other media. The urge to communicate complex ideas is one common thread that ties language in with the GLF, but an equally important one is the nuts-and-bolts of how these ideas get communicated: syntax.

A language isn't just a dictionary of words and their definitions. In order to be intelligible, those words must be strung together into phrases and sentences that follow grammatical rules. I'm not talking about the grammar that may have been rotely drilled into your head in school, or about the vacuous notion of "standard English". I mean the grammatical rules that you picked up automatically when first learning to talk, the ones which tell you that "I'm gonna drink some water" is ok, but "me water drink will" is not. It's something of a misnomer to call these patterns "rules", since they're not written down anywhere or enforced by an external authority, and they evolve over time. They're really a particular sort of social convention, where the members of society tacitly agree to follow most of the rules, most of the time.
Whatever you want to call them though, there's no question that humans internalize these rules, follow them fastidiously, and do it all automatically.

Music is similar. It doesn't have nouns or verbs, but just like language it starts with basic building blocks and strings them together according to structural rules. A chimpanzee with a piano will never make music. He won't play chord progressions, scales, harmonies or probably even hold a beat. Without these he'll only make noise. When you lay all the rules (of either music or language) out explicitly, as professional linguists and music theorists do, they turn out to be *extremely* complicated - much moreso than the "grammar" you may have learned in school. It takes years of study to master them, and it's not even clear that we've discovered them all! Yet somehow, none of that really matters to our brains. They process these rules under the hood, and they do it so effortlessly that no training is needed to learn a language (at least as a child) or to appreciate a good guitar solo.

The complexity of tallying pales in comparison to music or language. In tallying the rules are simple and clear, so much so that even a child could explain tallying in one sentence. It's still a syntax though, and the fact that it was invented right at the time of the GLF is tantalizing.

Let me back up for a minute and state all this a little more generally. Language, music and math all fall under the umbrella of "symbolic systems". A symbolic system has two parts: a basic set of symbols (like musical notes, words, tally marks, etc.), and a set of rules for combining them into grammatical patterns. The sentence "America is a country" and the equation "2+2=7" are both grammatical statements. The second statement happens to be incorrect, but it's still grammatical. Even saying that it's wrong requires it to be meaningful. Ungrammatical statements, like "country is America an" and "+/3=" are neither right nor wrong. They're just gibberish.

Humans, and apparently only humans, have a natural affinity for symbolic systems. They let us bring vast, remote, or precise ideas down to earth, in a format that our brains can easily work with. The evolutionary advantage of symbolic systems, language in particular, cannot be overstated. It let us develop oral traditions and histories. It allowed intricate communication over great distances. It let us organize complex societies. This tool, I believe, is what enabled us to migrate out of Africa, displacing our cousins the Neanderthals in the process, and to create the world we live in today.

In fairness I should note that many archaeologists question whether a "Great Leap Forward" happened at all. According to them, the reason that we see such a blooming of art around 40,000 B.C.E has nothing to do with advances in our cognition. It's just that the population exploded, so there were more people around to make art.

These researchers point to a number of apparent instances of symbol usage before the GLF, such as Blombos Cave near the southernmost tip of Africa. Blombos was occupied some 75 thousand years ago, and besides unusually advanced stone tools, it has shell beads and patterned carvings made in red ochre rock. Are these not symbols? Even more striking evidence can be found in Krapina, Croatia, where Neanderthals were burying their dead 130 thousand years ago.

Burial of the dead, religious rituals and artwork are examples of "symbolic behavior", where objects and activities are given a meaning that transcends the things themselves. Symbolic behavior is closely linked with symbolic systems, with the big difference being that symbolic behavior doesn't necessarily have grammar. The fact that even the Neanderthals used symbols is compelling. Could it be that symbolic systems are much older than 40,000 year, dating all the way back to the time of Homo Heidelbergensis?

Actually these findings are not surprising, even if we assume that the Great Leap Forward marked the beginning of modern language. Language isn't one monolithic entity. It's a massively intricate endeavor, with a lot of moving parts. There is the vocabulary itself, with all the different classes of words that exist and ways they can be modified. There is a host of grammatical rules for putting different kinds sentences together. There are also more subtle aspects, like the physiology of the larynx, the effects of intonation, and the surprisingly important role of cadence. It would be absurd to think that all of these parts just popped into existence at once, fully formed, during the GLF. On the contrary, we should expect that our pre-GLF ancestors had some sort of a symbolic proto-language. Maybe the Great Leap Forward put the finishing touches on language (in the next chapter I'll speculate on what those "finishing touches" may have been), but it was only the final piece in a puzzle that had been evolving for a long time.

Even today, modern humans use symbols all the time in a non-linguistic context. Wearing a Rolex to work or hanging a cross on the wall sends a very clear message, without any syntax required. It is very plausible that Neanderthals did things like that, and even that they spoke in simple sentences, without mastering the full complexity of modern speech.

The essence of human language isn't the vocabulary, which maps words to their meanings. It's the grammar, which lets us construct new meanings from old ones. The tallying that Hera practiced was a conceptual leap beyond anything in Blombos cave or any known Neanderthal burial. She didn't just invent a tool for expressing a number; she invented a systematic way to express *any* number. On the way to evolving a full-fledged language we should expect to see an archaeological record full of symbols. Before a pot comes to a full boil it will simmer, and there will be some bubbles.

Fast forward to modern times. Language and math in Western society have changed a lot from their original forms. In contrast to life during the paleolithic, we now write emails, send texts, and get exposed to foreign languages from all over the world. Instead of tally sticks, we do math with hindu-arabic numerals and supercomputers. Standing at this vantage point it's hard to picture what life was like for Hera's people back on the plains of Africa. However, there are still isolated tribes in the world, who have been out of touch with the West for thousands of years and who follow a hunter-gatherer lifestyle more similar to our ancient ancestors. They provide a window into what the world used to be like.

Today, if you follow the Amazon River up through its tributaries and into the deep jungles of Brazil, you will eventually come upon the Maici River in the heart of the continent. On its shores you will find a tribe of some 400 people who call themselves the "Hi'aiti'ihi" which roughly translates to the "straight ones" (other people are "crooked"). In the English-speaking world the Hi'aiti'ihi are known as the Piraha.

What is Math?

After Homo Sapiens migrated out of Africa, they spread Eastward throughout Asia, and about 13,000 years ago a small group crossed the Iberian land bridge into North America. Over the subsequent centuries their descendants spread southward, filling North and South America with hunter-gatherer tribes. Some of these tribes later evolved into large-scale civilizations like the Inca and the Maya. But nestled in the heart of the Amazon, the Maici river valley became a time capsule of the ancient hunter-gatherers.

If you visit the Piraha today, you will see t-shirts on the men, aluminum can pull-tabs strung into necklaces on the women, and the occasional machete - the products of very limited trade. But other than that they tenaciously stick to a way of life that, though now foreign to most of the world, has probably predominated for most of human history. They have no recognizable social hierarchy, and certainly no governmental structure. You don't really need one when your society is small enough that everybody knows everybody else. History as we know it, even oral history, is not recorded, since they simply aren't interested in anything beyond living memory. There's no need to store food or plow fields, because the jungle provides abundant game, fruits and vegetables year-round. The culture places no value on chastity. People take multiple naps during the day. And except for a little bit of trade, they don't want to change a thing. As I bustle back and forth between airports and offices during business travel, it's easy to sympathize.

The Piraha speak a language as unique as their culture. Unrelated to any other living tongue, Piraha has the fewest basic sounds of any spoken language (although depending on how you count, it could be tied with Hawiian and Rotokas, a language spoken in Papua New Guinea). There are no words for colors. Like a number of other languages with ancient roots, large parts of the vocabulary can be whistled rather than spoken.

Another feature of the Piraha language, which is shared with many other hunter-gatherers, is that it lacks numbers. There are two words for describing quantity, which can be translated as "one" and "two". But the same words can also be used to mean "few" and "many"; they are for describing relative amounts moreso than absolute quantities. If you want to say "three", then you're simply out of luck. It's not that the Piraha don't understand the notion of larger numbers; they just have so little occasion to use them that there was never any need to give them names. In essence, they never have numerical needs that fall outside of their intuitive scope.

This might seem shocking to you at first. It certainly did to me! How can you live without ever having large quantities in your life? It turns out it's not as hard as you might think. As a quick illustration, how many cousins do you have? If you come from a large family like I do, you probably don't know off the top of your head; you have to break them up into your mother and your father's side, then count and add. But somehow not knowing the head count never gets in the way at family gatherings. Every person has their own name, face and story, and that's the basis on which you interact with them. Counting is used for things that are more-or-less indistinguishable. You might have to count your cousins if you're buying food for a family gathering, but this is a very special circumstance, and even then just eyeballing it is likely to work fine.

In practice we rarely need large numbers in our daily lives, and when we do it's usually for something fairly modern like commerce or calculating gas mileage. For the Piraha, it was just never worth the trouble to devise, learn and teach a robust counting system.

There are ways in which the Piraha are unique even among hunter-gatherers. I'll talk more about that in the next chapter. But the lack of numbers is pretty typical. Most humans who have ever lived probably never learned to count. For a large portion of your life it's just not necessary.

A lack of explicit counting doesn't necessarily mean a lack of numbers; it just means a lack of *precise* numbers. I mentioned earlier that you can probably eyeball a pile of food to see whether there is too much, too little, or about enough for a family gathering, without ever doing a head count or measuring anything. We don't know exactly how the brain performs pseudo-calculations like this, but it is a completely different process from the counting and arithmetic you learned in school. This instinct for quantities is a distinct mental module, sometimes called the "number sense". Mathematics is what happens when you combine the raw number sense with a linguistic construct like counting. However, you can also - like the Piraha - just use the number sense on its own.

We're used to thinking of math as a learned, humans-only venture. While that's certainly true of language-based higher math, the number sense is not species-specific. On the contrary, research in recent decades has shown that the number sense is actually pretty generic among higher animals. Infants who are only a few days old have it. It's also built into apes, pigeons, rats - basically every animal that has been tested. Perhaps most surprising of all, the raw human number sense is not much better than anybody else's. If you'll indulge my interest in the history of psychology, I'd love to give some background on this!

Something like the number sense might seem like a pretty useful evolutionary adaptation, that it would be reasonable to expect among smart animals. While that conclusion turns out to be correct, it's not as obvious as you might think, and researchers were initially shocked to discover that the number sense exists and is so universal. For example, estimating amounts of food could be a separate mental module from estimating how many predators are chasing you, since there's never any need to compare them. That is, it's totally conceivable that an animal would see a pile of three apples and a pile of three walnuts as being the same size, but wouldn't draw a connection between those piles and a group of three saber-toothed tigers. Our perception of numbers could also conceivably be broken up by the senses, so that seeing three objects was totally different from hearing three sounds. The idea of using the same concept to keep track of what we see, what we hear, and even the passage of time is extremely abstract, and it's not obvious that humans, much less other animals, would come by it naturally.

For a long time most psychologists didn't believe it. They thought that humans were born without any built-in ideas of quantity, and it was only after growing up some and interacting with the world that we could eventually learn it.

The standard-bearer of this school of thought was Jean Piaget. Piaget taught that children are born as the proverbial "blank slates", learning machines with no preconceived notions about the world they've been born into or how to make sense of it. Beyond a few basic instincts like how to breastfeed, they start off by just mindlessly consuming raw input from their senses, like staring at a TV screen full of static. Eventually though they start to notice patterns. The world isn't just a jumble of light and sound; it is filled with physical objects, which move around and tend to maintain their shape. However, objects don't generally teleport or anything like that. Amazing! It is also at this stage that babies develop "object permanence", or the realization that an object can exist even though they aren't seeing it.

With the understanding of physical objects firmly in place, as a brick in the foundation of their cognition, the babies then go on to develop more abstract ideas.

They realize two objects can be similar or different in their shape, color, size, etc. After this they develop the idea that objects can be collected into groups, and that the group as a whole can have properties. Eventually, after painstakingly acquiring all these layers of abstract ideas, the child realizes that two *groups* of objects can be similar to each other, specifically by having the same number of elements. And thus the baby (really a child by this time) finally arrives at numbers, as a pinnacle in the pyramid of their cognitive development. Phew!

It's all hogwash. Babies aren't even close to blank slates. As it turns out, evolution has primed their brains to pick up on exactly the kinds of patterns that a homo sapiens is likely to encounter in life. Straight out of the womb babies will break up their visual field into solid, compact objects. They will expect these objects to move continuously through space, and they are baffled by optical illusions that make objects seem to do non-physical things, like disappearing or passing through each other. They listen closely for human speech, and quickly learn to pick out the basic sounds in the language spoken around them. Of course those sounds are different for each language, and babies have to figure out what the sounds for their particular language are after being born. However, they do know to look for them. Piaget was right that babies are learning machines, but they aren't capable of learning *anything*. They are specifically designed to learn the skills, facts and ideas that will make them cognitively normal adults. To that end they have a lot of preconceived notions, and it turns out that one of those notions is quantity.

Probably the most dramatic proof that humans have a built-in number sense came from the American psychologists Prentice Starkley, Elizabeth Spelke and Rochel Gelman. They put a baby between two screens, one of which showed 2 objects and the other of which showed 3 objects. At the same time there was a drum beating in the background, in groups of either two or three beats at a time. Initially the baby would ignore the sound and spend more time looking at the screen with three objects, since it was more complex. After a while though the baby would spend most of its time looking at whichever screen matched the number of drum beats it was hearing. The baby could tell that three sounds were "the same as" three objects! This experiment has been done with babies down to a few months old, all showing the same result. The only explanation that holds water is that babies understand numbers.

That discovery might sound very impressive at first glance. However, it turns out there are major limits to a baby's number sense. The experiment I just described used two and three objects, and showed that babies can "count" up to three. But if you use larger numbers, like comparing four to five, the effect starts to disappear. It seems babies can't count any higher: their number sense doesn't discriminate precisely above that level. The babies are certainly doing something interesting, but it's not counting as we know it.

The number sense actually comes in two flavors. One of them is used for dealing with large numbers in approximate ways - I'll deal with it in the next section. The other side of the number sense is what the babies were doing in the experiment I just described, where they "counted" exactly, but only up to small numbers. This is called "subitization".

Subitizing has nothing to do with normal counting. It's a deeply ingrained cognitive process that, as near as we can tell, happens in all higher animals, and is a critical component of the number sense. The underlying mechanism seems to be closely tied to our sense of sight; when our brain breaks up our visual field into distinct objects, it takes note of how many distinct objects were noticed.

By the time people grow up and learn language and arithmetic they have a number of different tools for solving the same problem, some built in and some learned. Thus, when you're studying the numerical abilities of adults it becomes very hard to tell where subitizing ends and counting begins. Even if you count a group of three objects by subitizing them you're likely to say "three" in your head; if somebody was scanning your brain at the time they would see both parts of it get activated. For practical purposes it's easy to see counting and subitizing as two extremes of the same thing. Nevertheless, even in adults there are telltale signs that subitization is at play as a distinct activity.

As a test you can run on yourself, try counting the bars in this picture:

What is Math?

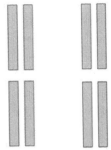

I'll wager that you didn't say to yourself "one two three four..." the way you were taught in school. Instead you probably broke the bars into pairs and counted by twos in your head: "two four six...". How did you know that each group of bars had exactly two? You subitized them! This is a hack that your brain does to make counting go quicker. Counting objects with words requires you to say all of the numbers in your head, which is a painfully slow process. Subitization on the other hand, being so deeply hardwired in our brains, is blazingly fast. By subitizing pairs of pairs and counting by two, you only have to say half as many numbers in your head. This hybrid approach combines the accuracy of actual counting with some of subitization's speed and efficiency.

Even though subitization starts to break down for quantities bigger than 3, evolution has not left us helpless. Above that point animal reasoning (it's the same for humans as for other animals, so there's no point in making an artificial distinction any longer) becomes more of a continuous approximation, called the Approximate Number System (ANS). This is what would allow you to eyeball the food at a family gathering and see that there's about enough.

The ANS doesn't recognize "sixteen" as a distinct quantity, but it can still recognize that a group of sixteen objects is more than a group of eight, less than a group of fifty and about the same as a group of seventeen. The ANS is a completely different mental module from subitization, and together they constitute our number sense.

In the same way that you unconsciously subitize objects even while you're counting with words, the ANS is always on, even when we're doing math a different way. Let's say I ask you to add 23 and 19 by hand, with carrying the 1 and all that. Before you even put pencil to paper, your ANS will recognize roughly how big 23 and 19 are, and will have some idea of what their sum should be. If you happen to botch the hand calculation and get something way off, like 23+19=200, the ANS will immediately recognize that something's wrong and sound the alarm. If you get an answer like 23+19=40, which is still wrong but very close to the right answer, your ANS won't be able to tell the difference. Whenever you solve an arithmetic problem the ANS is right there beside you. As you chug along with paper and pencil to get an exact answer, it marshals ancient neural circuits to do the same calculation in parallel, and it raises a red flag whenever the answers don't match up.

Personally, I like to imagine how the ANS works by picturing piles of apples. If you are asked to add the numbers 15 and 32, first imagine a pile of 15 apples and a pile of 32 apples. Then imagine combining those piles, and guestimating how big the total pile is.
 You won't get the correct answer of 47 exactly, but you'll get something in that range. It's a similar process if you're trying to compare two numbers to see which one is bigger: the ANS will convert both numbers into piles of apples, eyeball the piles, and then pronounce which one is bigger (or say that the piles are roughly the same).

Comparing the sizes of numbers is actually a great way to see how the ANS works hand-in-hand with counting in an adult mind.
 Researchers have done studies where an adult human is shown two numbers and has to decide, as quickly as possible, which number is larger. Answering the question correctly is of course trivial, but what's really interesting is how long it takes them to determine the right answer.

As it turns out, people give the right answer more quickly when they are comparing 7 to 3 than when they are comparing 57 to 53.

The difference in speed is small enough that test subjects didn't notice it, but psychologists have done a lot of carefully timed experiments, and its definitely there. How could this happen?

The absolute difference between 7 and 3 is the same as the difference between 57 and 53, namely 4. But 4 is a very big number relative to 3 and 7, whereas it's tiny compared to 57 and 53. The ANS is able to use the large *proportional* difference between 7 and 3 to immediately pronounce which one is bigger, without any need to resort to high-level brain functions. However, 57 is roughly the same size as 53, so the ANS realized that it is not up to the task. The baton is then passed to the slower, but more accurate, process of comparing the numbers' digits.

Going back to my earlier illustration, imagine comparing a pile of 7 apples to a pile of 3 apples. It's obvious which pile is larger - no counting needed. The same thing would happen if you were comparing 70 apples to 30, or 700 to 300. The actual numbers are unimportant, but the relative differences carry the day. However, if I ask you to compare a pile of 53 apples with a pile of 57 then it becomes a more marginal case. You'll have to count the apples, line them up side-by-side, or use some other mental crutch.

It is fascinating to see that the species which developed algebraic topology and put a person on the moon is, at the most fundamental level, no more numerically adept than a pigeon. But humans, alone in the animal kingdom as far as we can tell, aren't limited to our number sense. If a pigeon looks at a pile of 53 seeds and a pile of 57 seeds, it will never be able to tell which one is bigger; the pigeon has no way to go beyond its intuitive scope. Humans can't tell the difference either when they rely on their number sense, but we have another trick up our sleeve, a whole other cognitive module that can come to the rescue. We recruit our language faculties to pick up the slack when our number sense isn't up to the task. This is a much slower process, it's more cognitively taxing, and it's far less intuitive, but it gets the job done and gives the right answer.

In my mind this is the essence of mathematics: leveraging the power of language to bootstrap up from our basic mental tools, especially the number sense. Those cognitive tools, which together form intuitive scope, are a hard ceiling on the cognitive abilities of most animals. But for humans they are just a foundation.

In this book I try to make my case that mathematics is just a branch of language, but in all honesty that's not really the point of the book. I've spent most of my life doing math in one form or another, and I think there are a lot of misunderstandings that all types of people have about it. Many laymen think math is some sort of intimidating voodoo, that they aren't able to (and don't care to) understand. On the other hand, professional mathematicians often see it as some sort of transcendent truth, and fancy themselves to be explorers probing the deep mysteries of existence. And a lot of engineers don't really care. For them math is just a tool, and at the end a number pops out that they can use.

I think all these views are wrong. My real goal in this book is to show that math is a deeply human activity, an organic outgrowth of our minds that helps us make sense of all aspects of our experience. There's nothing mysterious or transcendent about it, and neither is it beyond anybody.

This chapter has focused on the simplest type of math possible: counting. Some people might argue that a book called "What is Math" should instead be talking about calculus, formal logic, topology, or something else that sounds fancy. But really counting is all you need to understand my core message, that math is just an unusual branch of language. Counting is as far beyond anything in the animal kingdom as Shakespeare is beyond bird calls, and for exactly the same reason; they are both instances of human language, the cognitive mojo that makes humanity so unique. The only reason counting doesn't seem "advanced" is that humans are wired to be so good at it, in the same way that we can speak a complex language like English so effortlessly. That ability is the real miracle, and the gap between counting and calculus is tiny by comparison.

The next chapter will talk a little more about language in general. This is a big enough subject to warrant a book in its own right, so I'll only focus on some of the startling ways that natural language is similar to math, both in how it's structured and in what it looks like in the human brain. The rest of the book will explore the parallels between math and language, and explain the history of why they came to be so different.

Above all, I will try to show what all this means from a practical perspective. Math is a powerful, important tool that is used - and mis-used - every day in our society. If we can look at math "as it is", then maybe we can use it more responsibly.

2 WHAT IS LANGUAGE?

A computer scientist is heading to store to pick up groceries. On the way out, she asks her husband what they need. He thinks about it for a moment, and then tells her "get a loaf of bread. Oh, and if they have eggs, get a dozen." She heads to the store, later returning with shopping bags stuffed with twelve loaves of bread and not an egg to be found. "What's up with this!?" the husband asks. "What on earth possessed you to buy so much bread?" The computer scientist looks at him with confusion. "They had eggs".

The point of this joke is only partly to poke fun computer scientists - it's mostly to highlight the ambiguities of language. A lot of grammarians would actually agree with the computer scientist on this one. If her husband wanted the eggs, then he should have specified: "if they have eggs, get a dozen *eggs*". As it is he only told her to get a dozen of *something*, and without that clarification the ambiguity resolves to the original thing being discussed. In this case that's loaves of bread. So the husband *had* accidentally told her to get the extra bread, and she was just taking him at his word.

Of course in context it's easy (at least for non-computer scientists) to understand what the husband actually meant. But that's just because we have a lot of extra knowledge that we bring to the table. First off, we know that you usually buy a dozen eggs at a time, and loaves of bread one by one. There's also a hint in the words we used; the word "dozen" is used almost exclusively to refer to eggs, even though it can technically be used for any group of twelve things. With all this extra information, your brain doesn't skip a beat in figuring out what the husband meant. When you read the joke, chances are you didn't even realize there was something amiss until the punchline.

As stated though, the husband was definitely being ambiguous. To show that this is the case, imagine instead that he had said

"Get a loaf of bread. Oh, and if they have a sale, get a dozen". Structurally this is the same sentence from earlier, just with "a sale" replacing "eggs". But now he's obviously obviously saying they should stock up on bread. The sentence is still ambiguous, but again it's context that has come to the rescue. We know that a sale is an event, and generally you can't "buy" one in any meaningful way. Furthermore, if there is indeed a sale on bread, then it might make sense to buy in unusual bulk. Counting bread by the dozen is still a little unusual, but since there are no eggs around anymore it's obvious what we're referring to.

In daily life context and literal meaning are generally aligned, so weird situations like this don't show up. Language is filled with statements that are ambiguous or even technically incorrect, but almost everything we say fits in pretty seamlessly with everybody's understanding of how the world works, so our pre-conceived notions can clear up ambiguities or mis-speakings. In other words, 90% of life falls well within everybody's intuitive scope In the unusual cases when we *do* say something truly bizarre, we're generally very explicit about it, leaving no possible room for ambiguity. It's a good system.

However, this system breaks down when we venture beyond daily life. If you need to interpret legal documents, follow an instruction manual, or understand the details of a scientific discovery, then it becomes a game of eliminating ambiguities. We shift away from our intuitive scope and rely more on precise, literal interpretation, because we're in a situation where our intuitions are not of much help. The meaning of a statement is no longer implied by the context; it must be nailed down by the syntax.

Recently, in my eternal quest for cat pictures and other diversions on the internet, I found a remake of the music video for Gangnam Style, which was put together by a bunch of MIT students.
    Students at MIT, I have found, don't do anything halfway, and this video was no exception. Besides going all over the Boston area for filming, they must have recruited hundreds of students and

faculty to re-create the giant dance sequences. I have to wonder: how many tests were flunked due to the man-hours spent on that project?

About three fourths of the way through the video, as the tempo is ramping up to a fever pitch, the loud music and crazy group dancing comes to an abrupt halt. The scene cuts over to an elderly gentleman sitting quietly in an office. He sips from a mug of tea, sets the cup gingerly on the table and looks calmly at the camera. He deadpans: "oppan Chomsky style". Immediately, the video cuts back to the crazy dancing and loud music.

The man in question was Noam Chomsky, a professor in MIT's linguistics department. There's a strong argument to be made that he's the greatest academic mind in the world today, and his works have been cited more than any other living scholar. In fact, in the all-time ranking of scholarly citations he is number 8, just behind Sigmund Freud. Chomsky is best known for being the father of modern linguistics, with a special focus on the ways that the mind processes grammar. He's not really interested in the fact that you buy eggs by the dozen, or the ways that such context can change the interpretation of a sentence. Instead he focuses on the underlying structure of language: the nouns, the verbs, the conjunctives, and the rules that string them all together. What are those rules, he asks? And more interestingly, what can those rules tell us about how our minds work?

Chomsky's academic work is pretty esoteric, but he is still famous outside the ivory tower. Not so much for his science, but for his politics.

Chomsky calls himself a liberal anarchist, but he usually gets lumped in with the far left of American politics. Not the left as it exists today, but more of the Marxist variety which has, eh, fallen out of favor. In his mind governmental authority is inherently illegitimate, and generally ends up functioning as a force for oppression. He vehemently opposed the Vietnam war, marching with protesters and even refusing to pay taxes that he felt were going toward genocide. In more modern times he has lambasted

American foreign policy, and raised a lot of eyebrows when he compared George W. Bush to Osama bin Laden. He has penned books against Israel, against American foreign policy, against the media, against large corporations... the list goes on. Now in his 80's, Chomsky is still one of the most vocal commentators in the world, and one of the best selling political authors.

Thankfully Chomsky managed to keep himself out of prison, and he has always carefully kept his radical politics separate from his first-rate scholarly work. He pioneered the idea of using mathematical models to understand a language's grammar, and was able to use this approach to show that, despite superficial differences, human languages are more alike than they are different. He also argued strongly for the idea that language isn't just a cultural artifact. It's a human-specific instinct that we are wired to pick up, almost like spiders and web-spinning.

Researchers in other disciplines have borrowed generously from Chomsky's work, and his ideas have popped up in psychology, computer theory and even evolutionary biology. Whatever you might think of Chomsky's ideas on politics, his ideas on science have changed the world.

I suspect that the reason Chomsky is best known for his politics is because almost nobody really understands his science. It's a little like Albert Einstein. Einstein is probably the most important physicist since Isaac Newton, but a lot of his greatest work was steeped in fancy mathematics that even very few physicists fully understand. So in response, he's mostly famous for being an eccentric genius with puffy white hair.

Some linguists focus on very accessible, intuitive topics. For example, cataloguing how the English language has evolved since the time of Shakespeare. Or maybe visiting exotic tribes and recording the dialect they speak. Chomsky, in contrast, is a pure theorist (again, just like Einstein). He doesn't perform any experiments or gather any data, and he doesn't work on intuitive things like words and pronunciations. His focus is under the hood, on the cognitive processes that underly our language instinct. For Chomsky, language is less an object of study in itself, and more of a window into the mysterious workings of the human mind.

If you've watched a child grow up then you've seen first-hand the almost miraculous way that they pick up their native tongue. It starts off with incoherent babbling. There are no words or meaning, but they still manage to mimic the cadence and sounds of the speakers around them. (It currently looks like the cadence is actually how newborns distinguish between speech and other sounds in the first place, kick-starting the whole language learning process.) Eventually a few words like "no" and "mama" manage to stick, becoming forever associated with concrete meanings. The process then takes off like wildfire, and at their peak children are picking up 10 to 20 new words per week. Grammar takes more time to develop, but it emerges in the same spontaneous way, evolving from short phrases into long, articular sentences.

As I mentioned in the last chapter, when I say "grammar" I don't mean the formal stuff that you may have been lambasted with in school. That is a different, less interesting, and largely misguided topic called "prescriptive grammar". Prescriptive grammarians are the ones who complain about split infinitives and decry perfectly good (albeit recently minted) words like "ain't" and "y'all". They are also the ones who will argue about the meaning of "if they have eggs, then get a dozen".

"Descriptive" linguistics, on the other hand, is a science in the truest sense. The goal is to study how homo sapiens communicate in the real world, not to lecture them on how they *should* do it. Grammar just means the conventions that we use for making intelligible statements. It is the difference between "I like to drink milk" and "like milked me drink". People of all cultures, dialects, races, educational levels and socioeconomic statuses adhere strictly to such conventions when communicating. All of these sets of rules are pretty much equal when it comes it their complexity, difficulty to learn, and their irrationality. The wonderful thing, the great miracle of linguistics, is that societies invent these conventions spontaneously, and individual humans learn and apply them by instinct.

This of course begs the question; what are those rules of grammar, and how do children learn them? In the interest of full disclosure I'll tell you up front: we don't know the answer yet. There are, however, some very exciting partial answers. I'll get to those in a minute, but first I want to give you some historical context and explain what the rules of grammar most definitely are *not*.

For a long time researchers thought that grammar - or at least, the process by which we learn it - wasn't based on abstract categories like nouns and verbs. It was about the words themselves, and which combinations are or are not allowed. Babies learn grammar by hearing enough adult speech that they figure out which combinations of words do and don't occur in practice.

A child will hear the phrase "the dog" many times while growing up, so their brain comes to recognize that the word "the" can be followed by "dog". But they will never hear "the" followed by "am", so they come to learn that "the am" is ungrammatical. They will also notice that the word "I" is often followed by the words "am" or "will", to make the grammatical phrases "I am" or "I will". However, the phrase "I is" never occurs, and is deemed to be ungrammatical. In this view learning language is all about context. Whether a given word is grammatical is determined by the words immediately surrounding it.

There's a mathematical framework that captures this view of language, called a Markov chain. A Markov chain has no notion of nouns, verbs, clauses or any other grammatical constructs. Instead it's giant lookup table, which says how likely every word is to be followed by every other word. A Markov chain might tell you that the word "I" is followed by the word "am" ten percent of the time, by "will" five percent of the time, and never by the word "is". The lookup table will cover every pair of words in the language. If some pair of words has probability zero, like "I is", then that pair is deemed grammatically incorrect.

A Markov chain describing a language like English will be staggeringly large, since it has to account for every possible pair of words, of which there are millions. However, the basic idea is conceptually simple. When you listen to somebody speaking, your brain is keeping track of every pair of words you hear, checking them against the Markov chain in your head to making sure that what's being said is grammatical. If you hear something that has zero probability of occurring, the "bad grammar" alarm goes off in your head.

If a vanilla Markov chain isn't sufficient for your linguistic needs (and it never is for an actual language), there are fancier versions available. For example, you could take more words of context into account. Maybe the combination "the dogs are" is allowed, but "eats dogs are" is forbidden. You can keep going like this, to two or three or four words of context, making a richer and more powerful picture of the language. But the cost of this approach is that there are many more word combinations to keep track of.

And that's all there is to it. Markov chains are just a way of modeling the power of context by storing all possible word combinations, of a given length, in a giant lookup table. The traditional rules of grammar are still there, but only indirectly. Adjectives like "brown" will have non-zero probability of being followed by nouns like "fox", but won't be followed by articles like "the" or "an". There are exceptions of course, like telling somebody to "brown the meat in the skillet", but by and large these patterns will prevail.

It might seem to you like this is a pretty fishy way to look at language, and if so you've got good instincts - we'll get to that in a second. However, for many real-world applications, Markov chains are fabulous! In datacenters around the world, supercomputers are fed enormous corpuses of text (the contents of the library of congress is on the small end) to determine which words follow which, and how often, in English and other languages. These industrial scale Markov chains always take more than one word of context into account - six or so is more typical - and employ sophisticated tricks to handle the incredible size of a Markov chain's lookup table.

Companies like Google and Microsoft use these Markov chains for all sorts of tasks that mimic human intelligence. For instance, your phone might autocorrect a word you're typing if it is highly unlikely given the context, and if another word is much more probable. I used to work at Google, and while there I used Markov chains to find reasonable replacements for missing words in printed documents. One of the biggest areas where they get used is in speech recognition, and they lie at the heart of tools like Siri. If a garbled word could either be "quick" or "quack", the computer looks at the surrounding words to figure out how likely each alternative is in context.

In my work I've used Markov chains in a variety of areas. In something like analyzing speech or decoding DNA there are always subtle patterns, rules and edge cases, most of which you don't know, and most of which you don't actually care about either. If all you want is a simple way to take context into account and figure out which words are likely, then Markov chains are a blunt instrument that does a reasonably good (and often an extremely good) job, without having to tease apart how a system actually works under the hood.

Here's the rub. Even if Markov chains are a useful model of language in practice, in principle they are fundamentally not enough. Human being are NOT carrying around giant lookup tables in their brains. Humans are doing something completely different, something that we don't fully understand yet, and Markov chains are just a brute force way to approximate it. It was Noam Chomsky who put the final nail in Markov's coffin, and he did it with five little words: "colorless green ideas sleep furiously".

"Colorless green ideas sleep furiously" is, of course, a ridiculous thing to say. It's also an impossible sentence if you look through the lens of Markov chains. Nothing is both green and colorless, so the "colorless green" part of the sentence will have probability zero. Ideas are never green - probability zero again. It's meaningless to suggest that ideas can sleep at all, let alone that they can sleep "furiously". The sentence is patently absurd on every level. But it is not gibberish; it conforms perfectly well to the rules of English grammar.

If you want proof, try this alternative sentence on for size: "colorless green ideas sleep furious". This one raises a red flag on a gut level that wasn't present earlier. Ideas may not be able to sleep "furiously", but saying that they can sleep "furious" is even more wrong. "Furious" isn't just the wrong word - it's the wrong *type* of word. If we want to break out the linguistic jargon, then "furious" is an adjective whereas "furiously" is an adverb, and the word "sleep" can only be modified by an adverb in this sentence. However, you don't need technical terminology to know that something is deeply wrong with the sentence. A combination of words can be statistically unlikely to occur because it's ridiculous to say, or it can be unlikely because it's grammatically wrong.
  "Colorless green ideas sleep furiously" shows that these are different things in our brains, but Markov chains lump them into the single metric of probability.

Chomsky pointed out another problem with Markov chains, simpler and even more damning than colorless green ideas. The biggest problem with Markov chains is simply the staggering size of the required lookup table; a few years of a child's life just isn't enough time to learn the whole thing. To be concrete, a mediocre English vocabulary contains about 2,000 words. This means there are 2,000 * 2,000 = four million possible pairs of words, and in order to train a Markov chain, the child needs to learn the probability for every pair. If a child heard one word per minute for 16 hours a day, it would take eleven years to make it through all of the four million pairs, and that's only with hearing each pair once! In practice it's far, far worse than that: any passable Markov chain would need more words of context, and the size of the table grows exponentially with the amount of context. There simply aren't enough hours in a child's life to learn a Markov chain for a language, but by a few years of age children are speaking in full, grammatically correct sentences. However it is that we learn grammar, evidently those few years are all it takes.

The best way to look at why Markov chains fail is by analogy. The four million word-pair probabilities acts as "knobs" that you tune paint a fine-grained picture of the language. You know the probability that the word "I" is followed by "am", that "red" is followed by "bulldogs", "Red" is followed by "Sox", and so on. It's like describing a photograph by saying what color every pixel is. Sure: if you know the color of every pixel then you can infer whatever you want to know about a photograph. But what you really want to know is "it's a picture of Grumpy Cat sitting on a chair". The pixels themselves are incidental, a means to an end. You don't care about pixel colors; you care about the high-level patterns in the picture.

One simple example of a high-level pattern in language is nouns. Every natural language on Earth has them, and from a grammatical perspective they are interchangeable; you can replace any noun with any other noun and still keep a sentence grammatically correct. The rules for how a noun goes into a sentence are very complicated. A noun like "dog" can do an action, as in "the dog chased the cat", or it can be the recipient of an action if the cat chased the dog. The dog might not directly take part in the action of the sentence at all, if "the cat chased the dog's ball". The rules may be very complex, but you only need to learn them once in order to apply them to every noun in the lexicon. If I tell you that the non-sense word "frunk" is a noun, then you automatically know that "the frunk is here" and "every frunk has its day" are grammatical sentences. You can even guess about how to change the word around, and say "I see several frunks" or "that's a very frunk-like house". Depending on what a frunk is these phrases could be as ridiculous as "colorless green ideas sleep furiously", but they're all clearly grammatical.

Before you call me on it, yes: there are exceptions to what I'm saying. The word "deer" is the plural of itself. For most words you indicate possession by adding a "'s", but the word "you" gets changed to "your". Many languages have another little oddity called "gender", which is very common but which English happens not to have. In Spanish every noun in the lexicon is arbitrarily assigned to "masculine" or "feminine", and adjectives get slightly modified according to the gender of the noun they describe. So two nouns are only interchangeable if you also go through and make sure that all the adjectives line up. Every natural language, in practice, has a bewildering complexity of exceptions and edge cases. It is designed haphazardly by a committee of thousands to millions of speakers, spread across time and space, all tweaking the language one sentence at a time.

There are two ways to respond to this complexity. On the one hand you can catalogue it. The plural of "millennium" is "millennia" because the word is derived from Latin, and that's how Latin pluralizes its nouns. Written German capitalizes all of its nouns, rather than just those at the start of a sentence, because Martin Luther liked to capitalize nouns that were religiously important wherever they occurred in a sentence, and the habit caught on. Cataloguing all these eccentricities highlights the way that language is influenced by culture, history, ideology and biology. It has even allowed us to reconstruct large parts of human history, by looking at how languages have evolved and rubbed off on each other.

The other approach is to look past these complexities at the underlying patterns. It's cool to know why "millennium" has an irregular pluralization, but it's much more interesting to realize that this is the exception that proves the rule. Every language has default rules for pluralization which it only violates in rare cases. The fact that such rules always exist is what's most exciting!

Chomsky's key idea was to codify all of these universal language properties, combining them into a so-called Universal Grammar (UG) that gives the structure of all the world's languages. Presumably, if some property is shared by everything from

English to Japanese to Navajo then it's not an accident of culture or history. It is there because of some deep reason that applies to all humans. Some common features are there for obvious reasons of human biology, like having words for "food" and "sex". That's not what I mean when talking about the UG. I mean structural similarities in the grammars, which point to the mechanisms in your head that deal with language.

Let me give you one example of such a non-obvious rule. In English you say "I ate with chopsticks", but the word ordering in Japanese is more like "I chopsticks with ate". There are two differences between these sentences. The first difference is the structure of the adverb phrase: English says "with chopsticks", putting the preposition before the noun, while "chopsticks with" puts the noun first. English likes to start its adverb phrases with the preposition in general, like if I went to "to Australia" or borrowed a book "from John". Japanese is equally devoted to its noun-first approach. The second difference is in the structure of the verb phrase as a whole. Do you put the adverb phrase "with chopsticks" before or after "ate"? English puts it after the verb, Japanese puts it before. Lots of languages follow the English convention for both of these questions. Lots of other languages instead follow the Japanese convention for both questions. But almost no languages mix and match; you simply don't see "I ate chopsticks with".

One tentative explanation for this is the idea that - at least by default - the same part of the mind processes both adverb phrases and verb phrases. It acts as a switch that gets flipped one way or the other as you pick up your native tongue. You could of course force yourself to mix and match the rules, but it's kind of unnatural. The path of least resistance is for a community of language speakers to flip the switch one way (or the other) and apply it consistently across all types of phrases. There are many other patterns like this, that give hints about the mental machinery of language.

Linguistics is a science, so by definition the UG is a work in progress. Edge cases are constantly being discovered, and reformulations are made to existing ideas. Hypotheses are tested and rejected, and of course there are exceptions to most rules. That being said though, we do have a good rough draft going. Markov chains were a promising approach at first, but in the end they weren't up to snuff. Chomsky needed a new model to capture the way human language actually works, and he found the answer (or at least, a much better one) in something called a context-free grammar.

A lot of people feel like language is an organic, creative activity that humans naturally take to. It is rich, accessible, and the room for nuance (inflection, tone, etc) is unlimited. Math, on the other hand, is all about rigid, arbitrary rules and abstractions. It is sterile and uncompromising - quite the opposite of language. Most of the time in this book I try to paint math in a gentler light, and show that it's a lot more like spoken language than most people realize.
  This chapter is different. If Markov chains seemed complicated and abstract to you, then I'm afraid that context-free grammars (CFGs) will be worse. Like it or not though, CFGs are the nuts-and-bolts of how language actually works. My point here isn't to shove linguistic formalism down your throat. The real goal is to show that, when you peel back the flesh of language, you find that the skeleton underneath is as rigid and alien as anything you'll see in a math class (and don't worry - after this chapter things will come back down to earth).

Looking at the syntax of a language as a skeleton is actually a great analogy. It's not a list of arbitrary constraints so much as a framework for building up ideas. We can only run, jump and dance because our bones and tendons strike a delicate balance.
  They are flexible enough to let us move around, but rigid enough to keep us from collapsing. In the same way, grammar is not a ceiling or wall so much as it's a scaffolding.

In the earlier discussion about Markov chains we showed conclusively that they can't be the whole story when it comes to a language's grammar. However, we didn't say much about the critical ingredients that they leave out. The big place where a Markov chain falls flat is that it only brings into account a finite amount of context, looking at a sentence strictly as two-word (or three-word, etc) phrases. This precludes any high-level structure to the sentence. For example take the following, which everyone would agree is grammatically correct :

> "If you have breakfast after you get up in the morning, then you should leave some for me".

The overall sentence is an "if-then" statement, and the word "then" only makes sense here because "if" came before it. However, the matching "if" and "then" are a full eleven words apart, which means a Markov chain would need eleven words of context to see whether a statement is grammatical. Let's say you feed this sentence into a Markov chain with only two words of context. It will work through the words one at a time, making sure that each word jives with the word immediately preceding it, and forgetting all words prior to that one. By the time the chain reaches "then", it will remember that the last word was "morning", but it won't remember whether "if" was present at the beginning. A two-word Markov chain isn't able to say whether this sentence checks out.

At first you might suggest that I use a fancier Markov chain, which takes in eleven words worth of context (never mind that such a big lookup table couldn't fit into the biggest data center in the world, let alone a human brain). That fixes the issue for this particular sentence, but it just kicks the can further down the road. I can still break the new Markov chain by adding more filler words, to make "if" and "then" as far apart as I wanted. "If you have a very healthy breakfast, which might include oatmeal or even flaxseed, after you get up in the morning, then...". I can cram in as much padding as I want, but the "if" and "then" are still inextricably linked.

Markov chains are like a man in the dark, groping to explore his immediate surroundings as he walks. If he has very long arms he might be able to feel things several feet in front or behind him, but ultimately he is limited by his arms' length. He can't just look forward or backward to see the bigger picture. This amnesia of Markov chains is central to their functioning. It's what makes them mathematically elegant, conceptually simple and computationally tractable. Unfortunately though, it is their Achilles' heel when it comes to natural language. Mother Nature isn't using a Markov chain.

Let's take a different perspective. The sentence above is structured as "if X, then Y". X and Y are allowed to be any complete statements, no matter how long they are. In the same way that you can swap out any noun for any other noun and still keep a grammatically correct sentence, you can swap out any complete statements in an if-then sentence. In linguistic jargon, X and Y are called "subordinate clauses", and they can in turn have their own subordinate parts. It's a hierarchical organization, and no matter how big X is, the "if" and "then" stay linked.

To show how this hierarchy works in the UG, let's go back to Chomsky's "colorless green ideas sleep furiously". This sentence isn't an if-then statement, but it has its own hierarchical structure that you can picture in what's called a "syntax tree":

The sentence as a whole has two parts: the noun phrase and the verb phrase. These contain the sentence's main noun "ideas" (aka the subject of the sentence) and the main verb "sleep" respectively, along with any modifiers to them. A noun phrase can be a single solitary noun, or it can be modified by adjectives like "colorless".
The noun phrase in this case has a nested structure, since you can always add another adjective to make a more complex noun phrase. The verb phrase can similarly be just a solitary verb, or it can have its own substructure.

The noun phrase and verb phrase in this sentence are both pretty simple (a noun with two adjectives, and a verb with an adverb respectively), but they could also be quite intricate. We could instead have said "colorless green ideas, which have a history of bad dreams, sleep furiously when they don't take their medication". The noun phrase in this case is "colorless green ideas, which have a history of bad dreams", and it's a whopper.
We aren't just pre-pending some adjectives. We are fleshing out "ideas" with supplementary information about their bad dreams.

This is a complete thought in its own right, which has been rebranded into a supporting role. Ideas not taking their medication is another complete thought, which has been embedded into the verb phrase. In principle these can have their own complete thoughts embedded into them, and so on ad infinitum.

You may have gone through sentence breakdowns like the one above when you were in school. If you did, then you probably remember how excruciatingly tedious they get, and how unwieldy the hierarchy can become even for seemingly simple sentences. It's exciting that sentences can be broken down this way (well, at least it is for me), but actually doing it is a royal pain. Somehow though, our minds do it automatically when reading or listening to somebody. It's the core of how language works!

Context-free grammars (CFGs) are the mathematical embodiment of hierarchical structure like this. We call them "context-free" because we look *only* at the way a sentence is structured. We don't care what the words actually mean. We also don't care whether a sentence is reasonable or ridiculous. But we do care, very much, about whether the words are nouns or adjectives, and whether every "then" has a corresponding "if". Language in the real world has been endlessly jury-rigged by cultural conventions and historical accidents, blurring its core structure beyond easy recognition. CFGs are a tool for extracting this underlying grammar out of a language and studying it in isolation.

What is a brain, that it can parse a CFG? This is the real crux of Chomsky's school of linguistics. Studying the brain itself has historically been a biology problem, and tells us more about how neurons fire than how those firings combine into ideas.
Neurobiology is a bit like trying to understand how a car runs by studying the metal it's made of. On the other end of the spectrum is psychology. It catalogues how humans act but doesn't give the mechanisms of why. Psychology is like studying cars by examining traffic patterns. CFGs though get right to the heart of the matter; they are figuring out how an internal combustion engine works. Parsing a CFG is information processing, a well-posed engineering problem, and the brain is Nature's solution to it.

How does your mind parse CFGs out of spoken sentences? How does it translate the CFGs of your own thoughts into sentences that you can speak? This is something that scientists can really sink their teeth into!

Before I dive into the formalism of CFGs, I should give one final disclaimer. Even in Chomskyan linguistics CFGs aren't the end-all-be-all of grammar. A CFG only gives the "deep structure" of a sentence, the way our minds think about it. Before a sentence is actually spoken though, it sometimes gets shifted around a little bit according to "transformation rules" which I won't get into. CFGs can't capture all of the transformation rules we observe in real speech, so there are other factors at play in grammar. But CFGs do form the core of it, and that's what we'll stick to in this chapter.

If you ever dig up some of Chomsky's linguistics work you will find that he uses a very dry, mathematical notation for describing syntax trees. It's the same ideas that I'm talking about here, but he describes them in a much more austere, precise way. It doesn't make for light reading, but it does let him be concise and exact - sort of like legalese on steroids - and this precision is crucial if you're trying to make a detailed scientific inquiry. Let me try and give you a flavor for how the more formal notation works. It may seem pretty weird at first, and you'd probably have to become a trained linguist to fully appreciate how useful the formal notation can be. Remember though: it's just another way of looking at syntax trees.

The easiest way to think of a CFG is as a game, sort of a rote logic game like Sudoku. You start off with a single symbol S, which will act as a place-holder to mean any complete statement. Later on in the game you will have a list of symbols, nested into a hierarchy with parentheses, and some of the symbols will be actual words rather than amorphous place-holders. At first though, you start off with just a solitary S.

At each step in the game you can swap out one symbol for a group of other symbols, and there are rules that describe which swappings are allowed. For example, there could be a rule that lets you replace an S with "if S then S". In this case you have two place-holders where there had been only one, and the sentence becomes more complex. Other times you can replace a place-holder with a single word, as in replacing N (short for noun) with "dog". A replacement like this reduces the total number of place-holders lying around. You win the game when all the place-holders are gone, and the sequence of words you have left is called "grammatically correct". In a practical CFG the place-holders will represent grammatical categories, like verbs or noun phrases, and swapping out symbols represents the syntax tree above. However, the CFG itself is just a meaningless game of swapping symbols.

Let's take a simple, specific example of a CFG that captures part of the English language. Besides the placeholder S for complete statements, it will also have placeholders for nouns, noun phrases, adjectives and adverbs. With further ado, here are the allowed substitution rules, along with what they mean in plain English

Rule	Meaning
S -> NP VP	a sentence can be a noun phrase followed by a verb phrase
NP -> ADJ NP	a noun phrase can be an adjective followed by another noun phrase
NP -> N	a noun phrase can be just a single noun
AJD -> {colorless, green}	"colorless" and "green" are adjectives
N -> ideas	"ideas" is a noun
VP -> V ADV	a verb phrase can have a verb and an adverb
V -> {sleep}	"sleep" is a verb

ADV -> {furiously}	"furiously" is an adverb

You can play out this game as follows:
```
S
NP  VP
ADJ NP  VP
colorless NP  VP
colorless ADJ N VP
colorless green N VP
colorless green ideas VP
colorless green ideas V ADV
colorless green ideas sleep ADV
colorless green ideas sleep furiously
```
This construction shows that Chomsky's famous sentence is grammatically correct.

Now let's go back to if-then statements, which are a particularly interesting type of rule. Don't worry: I won't go through a full gameplay again, but you can imagine how these new tricks would fit in. We saw above that "if X then Y" is a grammatically correct statement, and in CFG notation we write this as
```
S -> if S, then S
```
Rules like this one are perhaps the most significant ones in linguistics, because they introduce a whole new S into the expression. The new S can become its own little sub-sentence, with its own nouns and adjectives, and maybe even its own sub-sentences. The first few steps in a game could look like this:
```
S
if (S), then (S)
if (if S, then S) then (S)
if (if S, then S), then (if S, then S)
```
The complexity can bloom infinitely, and it's all grammatically correct English (even if it might be impossible to understand somebody who actually said the final sentence). There are many other of these S-generating rules, which let us say things like "I heard that S", or "when S, I will say that S". These S-generating

rules are called "recursive", and we will have much more to say about them later.

At this point a CFG might seem like a pretty sterile way to look at language. If that's how it feels to you... good!! That's the whole point. Language in the wild is such a complex, multifaceted activity that it's almost impossible to study any aspect of it in isolation. When we say that a CFG is "context-free" we mean that, as much as possible, we're stripping away all of our biases and preconceived notions. All that's left is the syntactic scaffolding. It's a little like studying lab mice that have been raised in a controlled environment for generations. Obviously there's a lot they can't teach you about what mice are like in nature, but the system is simplified enough that important features might jump out at you.

When I first encountered CFGs it was in the context of computers (CFGs are used for modeling a lot of things besides human language), and I'll admit that I thought they were dead boring. Now I'm older, and hopefully a little wiser, and I've come to recognize that CFGs are really a means to an end. They are the stuffy, formal embodiment of fascinating, organic processes that are going on in the real world. Though they might be tedious and unintuitive, CFGs are the key to understanding not just human language, but a range of other topics as well.

Actually, Chomsky wasn't the first person to use context-free grammars to study language. He certainly put them on the map in the western world, and he seems to be the first person to use them for studying how the human mind works. The essential idea though predates Chomsky by two thousand years.

In ancient India, a scholar named Panini set out to codify the Sanskrit language. Sanskrit was the language of religion and scholarship at the time, kind of like Latin became in the Europe, so it made sense to understand it from every possible angle. Panini pioneered the idea of studying a language's syntax independently from its meaning, and he invented what amounts to a CFG for Sanskrit. To give you an idea of how complicated language can be, Panini's grammar contained 3,959 separate rules - and as languages go, that's considered concise!

Now, nobody actually cares about writing down the complete grammar for a living language like English. As Panini's work suggests, it would be staggeringly complex, and it would have almost no actual applications. Even if you went to all the trouble your grammar would probably be out of date by the time you finished it, since languages are evolving constantly.

However, all of this is really beside the point. The goal of Chomsky and his ilk isn't to catalogue CFGs for different languages; it's to understand how our minds process grammar. We're trying to reverse-engineer the part of the mind that learns, parses and composes syntax trees. For example, a rule that is present in *every* language's CFG will probably tell us something universal about how the brain does language. Certain types of rules, it turns out, are easier for our brains to parse than others. Some types of recursive embeddings are used with reckless abandon in normal daily speech. For other types of recursive rules, the sentence becomes almost unintelligible after more than one layer of nesting. Cataloguing these observations amounts to getting performance specs on the mental machinery that processes language.

Let me come back to the notion of recursive rules, since they deserve very special attention. Note that in the CFG above, using a rule like
    N -> {ideas}
gets us closer to ending the game. This rule lets us replace a placeholder with a concrete word, and we win when all the placeholders are gone. We also get closer to ending when we use a rule like
    NP -> N

Doing this leaves us with the same number of place-holders, but it has nipped the NP in the bud. The NP could have grown into a noun phrase with many adjectives, but now its fate is sealed. The next time we touch this part of the sentence our only option will be to replace to N with a word.

In contrast, the rule
    S -> if (S) then (S)
is different. By using this rule, we replace the S with an expression that contains yet another copy of S (two of them in fact). The new S's will be complete sub-statements. This means that we could go on applying the same rule, in principle, forever, ballooning our sentence's complexity with every round of the game. There is a special word for rules that can spawn whole new sub-statements. They are called "recursive".

In general CFG terminology any symbol that can regenerate itself (directly or indirectly) is called recursive. For example, the rule
    NP -> ADJ NP
consumes an NP, yet gives as another NP we can still play with. When we are talking about language in particular, not the general theory of CFGs, "recursion" only refers to embedding complete statements within statements. So a recursive rule is one that can generate an S, embedding a complete statement within another statement.

Recursion is a central feature of language, because it lets us embed one complete idea within another. We need recursive rules to have if-then statements. We also need recursive rules in order to use clarifying statements, like "I will go to the hotel where we went on our honeymoon". The clause "we went on our honeymoon" could be a sentence all on its own, but in this case we're only using it to add more information to the word "hotel".

Recursion is the property of language that makes it infinitely expressive. If the CFG only had rules that brought you closer to finishing the game there would be only a finite number of ways the game could play out, and hence only a finite number of sentences you could compose. But recursion gives us the freedom to balloon our sentence out into arbitrarily many layers before we start getting rid of the place-holders. Chomsky knew that infinite expressiveness hinged on recursive rules. In his mind recursion wasn't just important; it was what made human communication qualitatively different from animals. The fact that we can handle recursion says there is something very special about the way our brains operate.

So, is a context-free grammar the way the human brain actually works? Well, partly yes and partly no. On the one hand, the core of the syntax for a language is pretty well described by a CFG. On the other hand, our brains are actually terrible at working with CFGs relative to, say, a computer. Humans are fabulous at parsing sentences with simple structure, but our ability falls off quickly. For some recursive rules, a sentence is almost unreadable when there's more than one layer of embedding. A computer, on the other hand, can happily parse CFGs 'til the cows come home, and there's no realistic limit to how much complexity they can handle.

So the grammar part of our minds must operate very differently from a computer. Recursion was probably added onto our brains as an evolutionary afterthought, jury-rigging hardware that wasn't really designed for the purpose. Our brains are the best syntax-parsers in the animal kingdom, but that's just because they're the only ones.

The place where our brains really excel isn't in syntax; it's in bringing context to bear. Humans are superb at combining several streams of data into one coherent message. This is why most people will buy one loaf of bread and a dozen eggs, even if "get a dozen" technically referred to the bread. When you listen to a sentence, one part of your mind is parsing its literal meaning and constructing a syntax tree. Another part of your mind is listening to the words themselves, making common sense guesses about what the sentence might be saying based on the things you're talking about. This part is all about context and pre-conceived notions. Yet another part of your mind is listening for tone and inflection, and gauging the body language of the speaker. All of these get taken into account by the listener, in order to decide what was actually meant.

Now that I've shown you how complicated everyday language can be under the hood, let's take a look at something that's actually a lot easier: math.

Arithmetic can be described by a CFG like English can. In fact, a *much* simpler one. Sentences in arithmetic are things like "2 + 3 = 5", or "2 < 10". The things they're talking about - the nouns in the sentences - are numbers and combinations of numbers, like "2 + 3". The statements they make are about whether two combinations of numbers are equal, less than, or greater than each other. If we use E as a place-holder for a combination of numbers (which is often called an "arithmetic expression", hence the E) then all of arithmetic boils down to this CFG:

    S -> E = E
    S -> E > E
    S -> E < E
    E -> (E + E)
    E -> (E - E)
    E -> (E * E)
    E -> (E / E)
    E -> {any number}

We could play the game this way (I'm being a little cavalier with my parentheses):

    S
    E = E
    E = E + E
    4 = E + E
    4 = 2 + E
    4 = 2 + 2

to get the statement 4=2+2.

Note that in the last step we could also have replaced the E with 3 rather than 2, and gotten the equation "4 = 2 + 3". This equation may be wrong, but at least it's grammatically correct. Something like "+=4/2", which our CFG can't produce, doesn't even have the dignity of being wrong. A CFG for English can also give us falsehoods like "all pumpkins are tangerines", and absurdities like "colorless green ideas sleep furiously". Remember that a CFG is context-free, which means that we don't get to opine about whether statements are true or false, good or bad, reasonable or ridiculous. All we care about is whether their structure checks out.

The CFG for arithmetic is recursive, because the symbol E can reproduce itself. We can do something like

    E
    E / E
    (E + E) / E
    (7 + E) / (E - E)

and balloon our statements to massive size, with expressions embedded in expressions, just like we did with natural language. It's not recursive in the linguistic sense, since you can't embed a whole S into an expression, but it's recursive from an information processing perspective.

In fact, recursion pervades all areas of math, even in something as simple as counting. If you don't believe this, bear with me for a minute. The number 77 is math's equivalent of a noun, but it isn't a stand-alone noun like "dog". Instead 77 is the number 76, embedded into an "add one" operation. Likewise, 76 is built up from 75, and so on down to small numbers like 1, 2 and 3. Tiny numbers are small enough that our brains can subitize them, and they count as stand-alone nouns in our numerical cognition, just like "dog". Any higher though, and counting becomes all about recursion.

We can even write a simple CFG that produces all counting numbers, expressing them using repeated addition:

    N -> 1
    N -> (N) + 1

You might (very reasonably) object to the nuisance of writing seven as $(((((1+1)+1)+1)+1)+1)+1$, as you would have to do with this CFG. However, it does serve to underscore the way that numbers are built off of one another. A sophisticated counting system becomes almost a language-with-a-language, with words for a core set of numbers and a syntax for constructing other numbers from them.

Thirty thousand years ago, when Hera was making notches in the Lebombo bone, she understood intuitively that each new notch was another layer of recursion, a way to bootstrap small numbers up into arbitrarily large ones. If Chomsky is correct, and recursion is indeed the definitive trait of human language, then the ability to cope with recursion is unique to humans. That would explain why other animals can't count, but Hera could.

Recursion is a hot topic, and we haven't seen the last of it in this book. We'll focus on it in the context of mathematics, but you should know that It pops up in a range of disciplines and has even been a hit in popular science. I've shown you recursion in the context of CFGs, but more generally it refers to things that refer or give rise to themselves. In computer science, any subroutine that calls itself is called "recursive". Many paradoxes in math and language are based on recursion, such as the phrase "this statement is false". Some people even make sweeping philosophical claims about recursion, suggesting that its at the core of our consciousness. "Self-awareness" is, after all, a thing being aware of itself, isn't it? Personally I think that claim is premature at best, and woo woo at worst. But however you cut it, recursion is a big deal.

What is the relationship between our words and our thoughts? This question has been asked for a long time, but mostly just by philosophers. In the modern day though, psychologists and linguists are also getting in on the action. Is it possible, for example, to reason logically without using words? If so then how much? Is language just a way to express ideas that we already have, or does it dictate the kinds of ideas we can have in the first place?

In the early twentieth century, a man named Benjamin Lee Whorf advanced an extreme answer to this question. He said that the language we speak lays the groundwork for our cognition, even with concepts as fundamental as the flow of time. He referred to this view as "linguistic relativity".

I would tell you that Whorf was a linguist, but that's not entirely true. He was actually a chemical engineer by trade, and his day job was working for an insurance company. He only moonlighted as a linguist, doing his own research, publishing papers and books, and even giving public lectures. You might (understandably) be raising an eyebrow at the idea of an insurance assessor styling himself as a scholar of language, and usually you'd be right. In this case, however, Whorf was not a quack. He was part of the rare breed of hobbyists who, with no university affiliation, still manage to become first rate academic researchers. He focused on Native American languages, cataloguing tongues like Hopi and Nahuatl. He even made major strides in deciphering Mayan hieroglyphics.

In the course of his work Whorf became convinced that certain native people, because of the language they grew up speaking, viewed the world in fundamentally different ways from Europeans. I don't mean that they had different opinions, values or world views. I mean that the most basic concepts and categories in their minds were different. Most famously he argued that the Hopi, whose language doesn't include a past tense, had a nonlinear understanding of the flow of time.

To put this in context, Whorf was a bit of an odd guy. He was a very competent linguist who did a lot of excellent researcher, but he also had an otherworldy fascination that sometimes bled into his scholarly work. He was very into what we would now call "new age" philosophies, like theosophy. To describe the impact of language on thought, he even coined the term "linguistic relativity" because he thought it was analogous to Albert Einstein's theory of relativity (I'm trained as a physicist, and I can tell you that that's baloney).

Not surprisingly, psychologists today agree that Whorf's more extreme views were bogus. Speaking a different language won't make you perceive time in a non-linear fashion, or open your eyes to other colors, or let you think in eleven dimensions. If those kinds of perceptions are what you're after, you'll do better consulting an LSD dealer than a linguist. Language codifies the basic ways that we perceive the world, but it doesn't create them.

However, that doesn't mean that language has no effect on how our minds work; there is a grain of truth to linguistic relativity. Most basically, speaking *some* language within the first ten years or so of life is a prerequisite for being able to learn a language at all, ever. History is dotted with tragic instances of children who grew up without exposure to adults speaking, either from being lost in the wilderness or situations of extreme neglect. When these children are eventually found and introduced back into society, it's often possible to socialize them and even teach them a large vocabulary. They can interact with people, and using words they can communicate basic messages. But they will never be able to put a grammatical sentence together, at least not one of any real complexity. The neural circuitry for processing syntax can only be laid down early in life. If that critical period is missed, then it's gone forever. However, if you learn to speak *some* language as a child, you will be able (maybe with difficulty) to pick up *any* language later on. And it doesn't really matter whether the first language is English, Inuit, or a tongue from Papua New Guinea. Any of them will do the trick.

There are also more subtle examples of linguistic relativity too. Take colors, for example. The cutoff between adjacent colors like red and violet is ultimately arbitrary, and different languages draw the line at different places in the colors spectrum. Some languages don't even have separate words for the different colors. As a native English speaker, my color perception is actually more sensitive near the cutoffs that English makes. I can look at two almost identical colors and tell you that one is purplish-red while the other is reddish-violet. But if you show the colors to somebody from a different language background, for whom both colors count as "red", you'll get a very different response. They are very likely to say not only that the two colors are both shades of red, but they are the *same* shade of red. They can't tell that there's a difference! An English-speaking brain is simply more discriminating in the part of the color spectrum where English draws its distinctions, and similarly for any other language. This effect is subtle, but researchers have shown that it's there.

There may be one other major example of linguistic relativity, a really fascinating one that ties directly into recursion and mathematics. I'm being very speculative here, and this speculation is based on some anthropology work that is wildly controversial in the halls of academia. But I just can't resist the temptation to go out on a limb. So while researchers battle it out in the halls of academia, lets go back to Brazil and up the Maici River, to the Piraha tribe we met last chapter.

The first person to really study the Piraha was a man named Daniel Everett. But Everett didn't go all the way up the Maici to study the language of an obscure tribe and learn their culture. At least not at first. Instead, he came to save their souls. He wasn't a researcher at all; he was a Christian missionary who happened to have a knack for languages.

Everett's path to becoming an academic linguist was unconventional, to say the least. He came from a blue collar background, and growing up was mostly interested in rock music. Then at age 17 he met a missionary couple in San Diego, and his life changed forever. He converted to Christianity, fell in love with and married the couple's daughter, and enrolled in the Moody Bible Institute. After completing his degree in Foreign Missions he entered the Summer Institute of Linguistics where he studied to become a missionary.

What is Math?

It quickly became apparent that Everett had an unusual talent for hard languages. So instead of a more normal project, he was sent on the most linguistically difficult mission available: a tiny tribe of about 300 people, nestled in the jungles of Brazil. The problem with ministering to the Piraha wasn't cultural resistance to change, fear of outsiders or anything like that. It was much more basic. Despite twenty years of work, no missionary had even been able to figure out their language. It was hoped that Everett's gift could scale this Tower of Babel, and they would finally be able to receive the Gospel. So in 1977, Everett moved with his wife and three children to Brazil.

Everett did manage to learn the Piraha language, and all subsequent work on the tribe has relied on his findings. However, from the church's perspective the project was an unmitigated flop. In the first place, the Piraha had zero interest in Jesus. Their culture is focused completely on the here-and-now. They don't even have creation myths or oral histories; anything outside of living memory is irrelevant. So it was baffling to them that Everett would care about somebody who had lived centuries ago in a far-away land. They just didn't get it.

To make matters worse, the Piraha are obsessively empirical, and disregard hearsay. They are so focused on hard evidence that it's even codified into their language. When you say something in Piraha, you add a suffix to the verb to indicate whether you heard that it was true, inferred it from other information, or saw it with your own eyes. There is no suffix for "people in my tribe have been re-asserting this for centuries, and I just accept it", so Everett was at pains to even explain how he himself had heard the Gospel. The Piraha were not impressed.

Now it's one thing for a missionary to fail in his mission. That happens. But it's another for the missionary to lose his own faith in the process. In the end it was the Piraha who ended up converting Everett. They converted him to their carefree way of life, their skepticism about hearsay, and their focus on living in the moment. They didn't need religion to live full and happy lives - quite to the contrary, they seemed to be the happiest people he had ever met - and Everett realized that he didn't either. He began having serious doubts in the early 80s, and by 1985 Everett had abandoned his faith and become an atheist. Of course, being a missionary deep in the Amazon jungle, with a devout wife and three kids in tow, is an awkward time to lose your faith, so he kept it under wraps for many years. When he finally revealed his atheism in the late 90s, his marriage dissolved and two of his children broke off contact. However, Everett thinks that the Piraha, and many other "primitive" people, have found the answers that many people in Western civilization have been grasping at for centuries, and which never seem to get any closer.

Getting back to linguistics, Everett discovered that there is a reason people before him had so much trouble with Piraha. I mentioned earlier that Piraha doesn't have words for numbers, but that's actually one of its more mundane features. The whole language is built out of a minuscule ten basic sounds ("phonemes"), tying it with Rotokas (a Hawaiian tongue) for the most spartan phoneme set known to science. For comparison, English has about 40 phonemes, depending on the dialect. Almost all of Piraha can also be whistled across long distances rather than spoken, which makes it excellent for coordinating a hunt in the dense forest. The same word is used for "father" and "mother", and there are no words for family more distant than parent/child and siblings.

All of these discoveries are unusual, but they are also thoroughly confirmed. Other researchers have verified Everett's findings independently. But Everett has gone one step further, and set off a firestorm in the research community. According to him, the Piraha language doesn't have recursion. There is simply no way to say "I hope that it rains tomorrow", or "the fish that I caught yesterday tasted good". Both of these sentences embed a whole thought, like there being rain tomorrow or you having caught a fish, into a larger one, and Piraha grammar has no recursive rules that allow such embeddings. In a few cases there are work-arounds, like the verb suffixes that say how you learned a statement was true. English usually conveys that information with embedding, but verb suffixes work equally well. There are only a few of these though, and each one can only work around one layer of recursion. According to Everett, Piraha is, in principle, a finite language.

It's well known that hunter-gatherers use recursive sentences less often than developed nations. That's not controversial. Even within a developed nation there is less recursion in the spoken language than the written. But the idea that recursion can't be used at all? Now that's incredible.

And yet, at the same time it could explain a lot. I mentioned that plenty of hunter-gatherer tribes don't have words for specific numbers, but they'll still do things like counting with their fingers or putting notches in sticks, in the same way that Hera did. They'll also have no trouble learning to count and do arithmetic in another language. None of this, however, applies to the Piraha. They have never been observed counting with anything, in any way. No tallies, fingers, or anything of the sort. Everett gave them math classes for months - at their request - and by the end not a single person could count to ten. Peter Gordon, a professor at Columbia University, followed up on Everett's work and concluded that the Piraha weren't just uninterested: they were cognitively incapable of learning to count.

Recall that the Piraha have no creation myths or oral histories. There is also nothing that we would recognize as art. No musical instruments. In other words, Piraha society looks an awful lot like humanity prior to the Great Leap Forward, which is when I suggested that modern language emerged.

It's known that there is a critical period for learning language. If you don't speak within the first ten years of life, you never will; the neural circuitry has to be laid down early. But there are many moving parts in human language, which don't always depend on each other. Maybe the critical period for learning a language isn't all-or-nothing, and you can miss out on certain language features if you aren't exposed to them. Specifically, could there be a critical period for using recursion, independently of the other features of a language? In more Chomskyan terms, maybe the recursive rules in a CFG are learned independently from the other rules, and have their own critical period for learning.

Math and linguistic recursion are structurally very similar: both recursive CFGs. So it's tantalizing to suggest that the same part of our mind processes both of them. If there is a critical period for general "recursive thought", and young Piraha miss it by learning a non-recursive language, then adult Piraha will never be able master any activity that demands recursive thinking. That includes counting and mastering foreign languages, and may indirectly include oral history and most art.

In the last chapter I suggested that language may have been the switched that flipped in the Great Leap Forward, giving rise to mathematics, music, cave paintings, and mythology. I'd like to refine that idea a little bit now. There are many different aspects of language, and surely they didn't all come about in the evolutionary blink of an eye. I suggest that prior to the Great Leap Forward people spoke languages very much like Piraha. But the Great Leap Forward is when recursion came on the scene, and with it a whole suite of new expressive power, ideas, and ways of describing the world. The emergence of recursion marked the beginning of modern language, as well as the beginning of true mathematics.

Now this is just a hypothesis, and obviously it would be unethical to test it by deliberately depriving children of recursive language to see if they could learn to count later. In fact, there is a very good counter-argument to be made to the "recursive critical period" hypothesis. The Piraha split off from other groups after humans had already populated the Americas, a mere 13 thousand years ago. If the Great Leap Forward was really the advent of recursion, then the ancestors of the Piraha spoke a recursive language. What would have made them stop? Well, let's suppose again that recursive thought has a critical period in early childhood. Suppose also that, by some cultural quirk or just random chance, recursion fell out of favor in a tiny, isolated tribe. This isn't entirely ridiculous; we already know hunter-gatherers don't use recursion as often as developed societies, so it's possible that one of them would have stopped it altogether. That tribe would then have a generation of non-recursive children. These children would later be incapable of passing recursion on to their own children, and so on. Recursion will have died out.

I should also caution that, even if this hypothesis is correct, it is a mistake to view the Piraha language as "primitive". They may lack recursion, counting, and even color words - conspicuously absent functionalities common in other languages. But their language has grown and evolved in other directions. Take, for example, the fact that it can be whistled over long distances. For the world in which they live that's much more useful than having subordinate clauses. Lacking recursion doesn't make their language "degenerate" any more than penguins are degenerate birds because their wings are no longer good for flying. To take an even more extreme example, many bacteria are arguably more "evolved" than human beings because they have developed radically new biochemistries, allowing them to live in extreme environments (high temperatures, high concentrations of salt, etc) that normal life couldn't tolerate. Humans, in contrast, have an off-the-shelf biochemistry; the only thing interesting about us is how we are wired together. Every language, like every species, has its bragging rights.

In arguing that math is just a branch of language, it would really strengthen my case to tell you that they happen in the same part of the brain. If arithmetic and geometry took place in the middle of the "language lobe" of your brain, then surely that would prove my case, right?

Alas it's not nearly that simple. Math and language are both complex activities with sub-skills spread across many parts of our gray matter. Natural language, for example, involves remembering the vocabulary you need, parsing the grammar of what other people are saying to you, and composing your own grammatically correct sentences. On top of that, you're also moving your mouth to form sounds and controlling the flow of breath over your vocal chords. That's just the language-specific part; you're also thinking

in a non-verbal way about whatever it is you're discussing.
Researchers have their work cut out for them in figuring out what parts of the brain do what! As a final sadistic twist from Mother Nature, it looks like brain layout also varies from person to person, so that even if we figured out which parts of your brain do what we couldn't necessarily generalize to other people. You know, in case the problem was too easy.

The study of how mental functions map to regions of the brain, which is called neuropsychology, has historically been built on the back of tragedy. Patients who have suffered brain-damage, from stroke or traumatic injury, often have extremely specific cognitive deficiencies which give a window into what exactly was taken away from them. More recently, we've been able to augment this data with fMRI machines, which give us a real-time view of the working brain. We are very far from a complete understanding, and the picture that has emerged so far is very complicated. It turns out that what people thought were "elementary" cognitive abilities end up being spread across multiple regions of the brain, and many parts of the brain seem to have several different functions. There isn't room in this chapter, or even this book, to do justice to what is being discovered, but I would like to say at least a little bit about the part that's relevant to math and language.

Touch your left temple, behind and above your eye. That part of your brain is called Broca's Area, and if you were to damage it you would get a condition called Broca's Aphasia. Patients with Broca's aphasia have normal vocabulary. They understand other peoples' words perfectly well, and have no trouble remembering the words that they want to say. However, depending on the extent of the damage, their grasp of syntax is lost. If you told them that "the man bit the dog" they would know that you're talking about a man and a dog, and understand that biting occurred. But they couldn't use the structure of the sentence to infer who bit whom. If the phrase came up in daily life they would automatically assume the dog was doing the biting - the exact opposite of what was meant. Aphasiacs understand language pretty well when other people are speaking, but that's just because so much can be inferred from context and the words used. The aphasiac isn't actually parsing the grammar. As soon as they have to put together their own sentences you'll see the problem. If you ask them to describe their day, they might tell you things like "awake", "coffee", "work" - things that they ate or did - but the ideas won't be knitted into coherent statements. Broca's aphasia is a tragic condition. The sufferers are perfectly intelligent individuals, who know exactly the point they want to get across and usually the words needed to describe it. Yet their brains simply refuse to translate the relationships between these words into the grammar of a sentence.

As I write this, it's reminding me of how lucky I am that my Broca's area is in good working order. Even sitting here writing a book about language, I couldn't fully explain to you why the sentence you're reading right now is grammatically correct. But even as I'm apologizing for my ignorance, I'm doing it in perfectly understandable English. I know grammar on an intuitive level, and my brain turns my thoughts into sentences on its own. I practically speak on autopilot! This is the work of Broca's area, which has quietly mastered English syntax and is quietly chugging along under the hood, freeing my conscious mind to focus on what I want to say rather than how to put it into words. Without the help of this autopilot I could still get simple points across, but it would be tedious, ugly and error-prone. The fluidity of speech would be gone.

A companion to Broca's area is Wernicke's area. They are both shown in the figure below. Broca's area parses the grammar of sentences independent of the words being used, but Wernicke's area focuses on the meaning of the words themselves. Broca's aphasia leaves people able to understand language from context, but unable to produce grammatical sentences themselves. People with damage to Wernicke's area, on the other hand, speak with normal syntax and cadence. However, they swap words out for each other randomly, and even make up nonsense words on the fly.

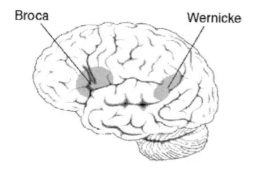

Interestingly, Broca's aphasiacs can generally remember pieces of normal speech, if they have them memorized. If they learned a prayer as a child they can recite it perfectly, but they can't re-phrase the prayer for you. This is because we memorize speech as just a sequence of raw sounds; we can re-construct the syntax on the fly later, but that's not how our brains store the information.

The same is true of math facts. An aphasiac can still recite "one two three four five six...", since it's been drilled into their heads since childhood, but counting backwards is a problem. They might also remember their multiplication tables, but will be unable to do a longer calculation.

Broca's area doesn't just handle the syntax for language and arithmetic. It turns out that it processes musical syntax as well. If you haven't had formal training in music then you might not think of it as having a "syntax", but it's there. Just as a native speaker can tell the difference between intelligible language and gibberish, no training is required to tell the difference between music and random noise. If a guitar player hits a conspicuously wrong note in the middle of a piece, even a piece you've never heard before, you know it automatically. That's because there is some pattern that your brain expected would be followed, and the wrong note broke this unspoken rule. The sum of all these patterns constitutes a "musical syntax", which has analogs of words, phrases, sentences and even paragraphs.

If you want to see an example of this yourself, take a song that you know (preferably one that doesn't include words, so the example doesn't get tainted with spoken language) and whistle the end of it out loud. Now whistle the end again, but leave out the last sound; end the song one note early. Irritating as heck, isn't it? It isn't just because you're used to how that song is supposed to end: dropping the last note from almost any song would be just as grating. This is because the rules of musical syntax put some constraints on what note a song can end on. If you cut a piece of music off randomly in the middle, then chances are that the new ending you've created isn't a "grammatical" one. This isn't a book about musicology, but there are lots of other rules like this, and usually we are as oblivious to them as we are to the rules of English. Oblivious, that is, until the instant the rules get violated.

MRI studies have shown that musical syntax, like linguistic syntax, is processed in Broca's area. The phrase "I were walking" and the hitting a wrong music note both set off warning bells in this brain region. Language and music generally use very different parts of our mental toolkit, but both of them have Broca's area running in the background, quietly making sure that the syntax all checks out.

The relationship between music and other mental faculties has struck a chord in the public. Some studies have suggested that listening to classical music can give an IQ boost, and people have made a lot of money selling products based on the so-called "Mozart effect". It looks like this is all snake oil - the IQ boost is small and temporary at best - but the fact that it caught on says something about how we view music. Anecdotally, many great thinkers have used music to calm their mind and stir their creative energies. Einstein, for example, liked to play the violin when he was stuck on a problem.

Math has its own syntactical rules, and they are no more complicated than those of music or English. In fact they're generally simpler, as we started seeing with CFGs. The difference is that Broca's area doesn't handle algebraic topology as naturally as it handles English or Beethoven, so we have to do some of the heavy lifting ourselves, on a conscious level. Broca's area lends a hand with foundational skills like counting, but it can't single-handedly do all the grammar-parsing, and this is part of why math can be difficult sometimes. With training we can learn the syntax of high-level math, but it doesn't come as naturally as spoken language, and as far as I know the jury is out on how much Broca's area is involved.

In the last chapter I discussed the "number sense", the built-in sense of quantity that humans share with other animals. We can instinctively recognize small numbers, like a group of three objects, by "subitizing" the things we see. For larger numbers the precision breaks down, and we do an approximate counting that you might call "eye-balling". The number sense resides mostly in a part of the brain called the intraparietal sulcus (IPS). It's located toward the back of your head, about a third of the way up your skull. The IPS is not related to Broca's area. The number sense has nothing at all to do with language.

Counting, though, is a different story. When you're counting or doing arithmetic your brain is operating on two parallel tracks.
  Your language faculty, including Broca's area, handles the syntax of how to write down numbers or do long division. At the same time, the IPS is going through the same calculations in parallel, solving the problems in a more natural (but less precise) way. If you make a major arithmetic mistake, the IPS will recognize that your answer is ridiculous and make you redo the calculation.
  Broca's area supplies the precision, while the IPS acts as a "sanity check".

This harmony between pure language and intuitions is the key to learning math, and to using it responsibly. Its the same balance that made the joke at the start of this chapter funny. You read "Get a loaf of bread. Oh, and if they have eggs, get a dozen." on several levels. Your Broca's area narrowed the sentence down to two possible interpretations. Another of your brain took the context into account, knowing what you're likely to buy at a grocery store. A third part remembered that "dozen" usually refers to eggs. You then seamlessly combine these different interpretations into a meaning.

It's tempting to say that, from a cognitive perspective, mathematics is built up on a foundation of our number sense, but I think that's only part of the picture. There's no question that arithmetic is based on our notions of quantity; it's the linguistic embodiment of our number sense. But geometry, in contrast, seems to be based more on our instincts about space and distance. Spatial cognition is spread throughout the parietal lobe of the brain, of which the IPS is only a small part. Most research on math in the brain has focused on numbers and arithmetic, but I strongly suspect that the IPS isn't particularly involved in, say, geometric reasoning. If we gave people in an fMRI geometry problems instead of arithmetic, I suspect that the IPS would be relatively quiet, but the rest of the parietal lobe would light up like a Christmas tree. There are other, more abstract areas of math that don't have clear ties with either quantity or space, and I expect they rest on still other basic faculties.

Speaking anecdotally, as somebody who has spent a lot of my life doing math of various sorts, there are several types of mathematicians, who work on very different areas of math, and who do it in very different ways. These differences could correspond to which underlying cognitive faculties people are best at. For myself, I think about math in terms of heuristic pictures, like you might see scrawled on a blackboard. This approach has served me very well in subjects like calculus and statistics.
However, I'm terrible at things like formal logic, and any skill I have in algebra comes from practice (I was a physics major - you do a LOT of algebra) rather than native ability.

My friend Ian is exactly the opposite. He's a mathematician of a very different stripe. For him it's all about abstract relationships, with no pictures to be found. Back in college Ian would pull easy A's in classes that I struggled just to pass, but it was was just the opposite in other subjects. We never tried it, but if you gave both of us IQ tests across different areas of cognition, and compared the results, I think it would be obvious where our differences came from.

The one thing that unites all the different areas of math is that, no matter which cognitive processes are doing the heavy lifting in your head, it ultimately gets translated into language when you put it on paper. This is exactly how normal language works; the ideas in your head aren't in words and sentences, but that's how they come out in conversation.

The difference between math and language is one of degree. In daily conversation, language is usually just a medium for getting your point across. Your understanding of what you're saying is essentially non-linguistic. But with math we're dealing with concepts that we have a more tenuous grasp of. So language isn't just a tool for communicating our ideas; it's also a crutch to help our reasoning.

Ok, now that you know what a context-free grammar is and what Broca's area does, let's back off from the theory a little bit. This isn't a book about cognitive linguistics; it's about how math fits into the human story. The next chapter talks about the early development of math. The agricultural revolution, the creation of large cities, and the founding of modern society pushed our brains beyond anything they had ever done out on the savannas where we evolved. Language was the tool we used to solve this problem, and that's when math split off from the language family tree.

3 THE RISE OF NUMBERS

In the first chapter of this book I took some literary license, describing Hera as the person who carved the Lebombo bone in ancient Africa. I made her up, of course, but we know that *somebody* carved the Lebombo bone, and we can infer a lot about that somebody from what we've found in Border Cave. We know what they ate, part of how they dressed, what kinds of tools they used. Add in a name, a gender, and an educated guess about what she was counting, and you have Hera. If you'll bear with me, I'd like to flesh out another historical character, who again we technically only know from the artifact he left behind. In this case, the artifact is the Moscow Mathematical Papyrus, which was pulled out of the Egyptian sand in the late 1800s.

The author of the papyrus was named Ram, and he was a scholar in ancient Egypt, around the year 1850 B.C.E. I say he lived in "ancient" Egypt, but by that time Egypt was already a very old civilization. The Sphinx had stood guard out in the desert for centuries, and the Pharaohs had been ruling the land for over a millennium.

Ram lived close to the Nile river, either in a hut or an open-air villa. The house was probably made of sun-dried bricks, but it's possible that he was rich enough to afford limestone. As far as clothing, Ram had to dress for the desert heat. Generally this meant a light skirt, with a belt around the waist and maybe some decorative pleating in the front. No shirt required. No shoes either, except for sandals on special occasions. But that doesn't mean Ram was dressed shabbily! Quite the opposite, in fact. He wore jewelry, and lots of it. Earrings, bracelets, arm bands - as much bling as he could get his hands on. He wore makeup too, including eye shadow and elaborate tattoos. And if he could afford it, he probably also wore a fancy wig.

Ram worked for the government, probably as an intellectual jack-of-all-trades. In a time when reading and writing were rare skills, a lot of his work was probably being a scribe. Simply writing a letter was a specialized task, so somebody with Ram's literacy skills would be in high demand. Which, I guess, means he could probably afford that wig...

However, Ram was more than just literate. Within the realm of scholarship, Ram's specialty was in calculations. If you wanted to know how many bricks were needed to build the new temple, how much grain a large field would produce in a season, or what the fair price for 150 cattle was, then Ram was the guy to talk to. Armed with pen and papyrus, he could chug out addition, multiplication, and even division. Big or small, the numbers didn't matter; his work was fast, legible, and above all correct.
This feat was a lot more impressive than it might sound. Unlike today, there was absolutely no recourse to calculators or computers. Everything had to be done by hand and checked by hand. There weren't modern pencils or paper, so Ram made do with reeds and papyrus (and likely something like an abacus). And bear in mind that Ram's job wasn't just calculating how many bricks you need to build a dinky little hut; the Pharaohs had a taste for pyramids.

As if that's not enough, Ram didn't even have a decent number system to work with. The way the Egyptians wrote numbers was conceptually a lot simpler than what we use today, and probably much easier to learn, but much more tedious to do actual work with. They borrowed one of their hieroglyphs to mean the number "one", another hieroglyph to mean "ten", another for "one hundred", and so on. Some of them are shown here:

What is Math?

Decimal Number	Egyptian Symbol	What it is supposed to be
1 =	|	staff
10 =	∩	heel bone
100 =	๑	coil of rope
1000 =	⚇	lotus flower
10,000 =	⌒	pointing finger
100,000 =	⌒	tadpole
1,000,000 =	⚇	astonished man

[Credit: https://www.ncetm.org.uk/resources/13153, http://www.math.wichita.edu/history/topics/num-sys.html#egypt]

If you wanted to write down the number 27, you would write the symbol for ten twice, and the symbol for one seven times, like this:

This is called a "sign-value" system, and it's very intuitively appealing. A sign-value system is reminiscent of how we use coins of different denominations in modern money, but it has the added simplicity of being base-10.

Don't be deceived though. After you've learned the number system you have to actually use it. Simply writing the number twenty seven in Egyptian numerals requires nine symbols, whereas our Hindu-arabic numerals have just two. This means much more writing to do for a given computation, and many more places where you could make a mistake. For mathematicians like Ram, doing hours of arithmetic by hand would have meant hours on end of pure tedium.

But Ram was more than just a human calculator. Even though he clocked a lot of hours chugging through rote arithmetic, his real value-added was in higher-level problem solving. People didn't come to Ram with number to add or subtract; they came to him with story problems. They had blueprints of the house they were building, or a map of the land they were farming, and it was Ram's job to translate these real-world problems into calculations. At the height of his career, Ram probably delegated the actual arithmetic to assistants.

Demand for Ram's services came from a lot of different sectors, from architecture to agriculture to accounting. But there was one area that stood out among them all, and was Ram's single biggest source of business: beer. The connection between math and alcohol long predates the American college scene.

We can safely assume that beer had the same recreational uses in ancient Egypt that it enjoys today, but it went well beyond that (the intoxicant of choice was actually opium, but that's another story). There was no refrigeration available, and Egypt was a hot place, so alcohol was pretty much the only thing you could store and drink without getting dysentery. Beer became so central to Egyptian life that even their gods developed a taste for it, demanding fermented sacrifices in exchange for divine favor.

The problem with brewing beer is getting the proportions right. How much strong beer can you make, given a certain amount of grain? How much hops should you put in? What if you like a lighter beer? These questions become especially important if the god Horus wants barrels of exactly the same beer that he's used to, and will punish the wrong sacrifice with a bad harvest or a raid by the Nubians. This is where Ram came in. He would calculate how much grain was needed to brew the right sacrifice. Or the reverse - he would tell you how much much beer you could make if you had one heqat of grain. Rather than beer, some of the grain could also be turned into bread, so Ram would calculate the tradeoffs of bread against beer.

I visited Egypt recently, and can personally testify to the prowess of Ram and his successors. After thousands of years tweaking the recipe, Egyptian beer is absolutely delicious! Sadly though, I cannot say the same for the wine...

At some point Ram decided to pass down his wisdom, and that's how we came to know of him. He wrote what's now known as the Moscow Mathematical Papyrus, one of the oldest math textbooks in the world. The papyrus was discovered by archaeologists in the late 1800's, and currently resides in the Pushkin Museum of Fine Arts in (you guessed it) Moscow.

The most interesting thing about the Moscow Papyrus is the way Ram organized it. Modern textbooks usually take a "math first" approach, where you learn the mathematics in the abstract and then apply it to (supposedly) real-world problems. The actual emphasis is on the math itself: technical definitions, logical proofs and memorizing algorithms. In some cases it's painfully obvious that the authors don't really care about applications; they just made up some half-baked word problems because you can't sell a textbook without paying lip-service to applications. But the Moscow Papyrus takes an "applications first" approach, and math is just a tool for solving real-world problems.

Apparently Ram didn't care about defining his terms in a rigorous way or showing logical proofs. That's not the way he did math. In fact, he probably wasn't even aware that you *could* do math that way - it would be another millennium until proofs really came into vogue in Greece. Instead, Ram picked 25 problems that were typical of Egyptian mathematics. Most of them were real-world story problems, patterned after the ones that Ram solved for a living. Ten of the problems, fully 40% of the papyrus, are about brewing beer, baking bread, and the conversions between them. I told you that beer was a big part of life! The other story problems covered issues like building ships or estimating labor schedules. "If it takes two workers three days to build one hut, how many days will five workers need to build ten huts?" That kind of thing.

For every problem Ram wrote out a detailed calculation of the answer, showing where you would change the numbers to solve a related problem. And that was it. The whole book is just a collection of worked examples. No philosophizing, definitions, derivations, theorems or proofs. It's a mathematical cookbook.

A textbook that consists entirely of worked examples might seem a little odd in the modern era, but in the ancient world it appears to have been almost universal. Other Egyptian papyruses have been pulled out of the sand, and they all take the same approach. The same thing applies to other cultures besides Egypt, even as far away as China. They might have more or fewer problems, covering different techniques and applications, and using different notations, but they are all cookbooks.

Besides this striking fact, there is one other nugget in the Moscow Papyrus that jumps out to the modern eye. In the tenth problem Ram calculates the surface area of a sphere. The correct formula for this area uses the famed number pi, which is the ratio of a circle's circumference to its diameter.

Pi is probably the most famous number in the world. It's approximately equal to 3.14, but technically its digits go on forever without repeating (3.14159265...). Students first learn about it in geometry class while studying circles, but it turns out that pi pops up throughout mathematics, in far-flung areas that you would never expect it. It threads through geometry, number theory, fourier analysis and others, knitting disparate parts of math into a single unified subject. Pi is a complicated and storied number, with many subtle theoretical properties. Many popular books and theoretical treatises have been written on this one number, and people have gone crazy obsessing over it.

But for Ram, pi was just 256/81. End of story. No decimal expansions that go on forever without repeating. No stories of mathematicians going crazy over it. No waxing philosophical about how it combines disparate fields of mathematics in an elegant, unified something-or-other. In fact, besides missing out on all the beauty and philosophy, Ram even got the number wrong! 256/81 is closer to 3.16 than 3.14, a distinction which seems to have gone unnoticed.

How could one of Egypt's leading scholars be wrong about something so fundamental as the freakin' value of pi? It's common for modern mathematicians to criticize Ram, or even to cite this as an example of how barbaric mathematics used to be, before we all learned better. But the fact is that Ram probably wouldn't have cared even if he had known about the error. 3.16 was good enough for his purposes. The economy was humming along, pyramids weren't toppling over, and above all the beer tasted good. What's to complain about? Besides, nobody's ever going to manufacture something that's perfectly spherical anyway, so why split hairs? A difference of 3.16 versus 3.14 - which is less than 1% - is the least of concerns.

This is a fundamental difference in paradigms. Today we look at math as abstract truth, that's only approximated by the real world. The value of pi isn't something that we go out and measure or test. It's something that we derive by careful reasoning and calculation. But for Ram, the only "truth" was the world around him, and the value of pi was an empirical fact. Generations of Egyptian scholars had found that, in practice, 256/81 gave good results. What's more, it was a simple ratio that made those grueling by-hand calculations much simpler to manage. Without question 256/81 was the thing to use. Ram was only "wrong" by modern standards.

A lot of people think the sister discipline of mathematics is science. However, the best candidate for that role is actually something a little less sexy; accounting. We didn't invent number systems and arithmetic to calculate the motions of the planets. We did it to keep track of flocks, store grain, conduct commerce and (sigh) levy and pay taxes. Those needs only came up when we moved from hunter-gatherer societies like Hera's into agrarian civilizations, like Ram's Egypt.

If you were a hungry caveman, you would go out into the words and forage until you were sated. Getting food for a hunter-gatherer is a complex task in a lot of ways. You have to remember the best places to find food, know what's safe to eat, have the strategy and agility to hunt an animal, the dexterity to sharpen a spear, and so on. However, none of it is exactly numerically challenging. If you lived in ancient Egypt though, these skills become less important than numerical talents. To feed yourself you would go to the local market and haggle on prices, understand the conversion rates between different good and knowing how to operate within your budget. Procuring meat didn't require hunting strategy or clever weapons, but it did require the ability to keep track of how many goats were in your flock. Plotting fields, seeing whether there is enough food to last the winter - all of this requires calculation, and our hunter-gatherer brains aren't natively up to the task.

This holds true up through the present day. A lot of the recent (at least, within the last few centuries) developments in mathematics have been driven by science, in particular physics, but 90% of the math that actually gets done is still about economics and inventories. In your personal life this means putting together a household budget, seeing whether there's enough gas in your car to make it to the next city, and making sure you bought enough carpet to remodel the living room. None of this goes beyond simple arithmetic and geometry, but these kinds of problems are the bread and butter of why humans use numbers.

I suggested in the first and second chapters that the Great Leap Forward - the flowering of culture that happened between 50 and 30 thousand years ago - happened because we developed modern language, specifically the ability to embed one idea into another. After this, humans spread throughout the world as a species of hunter-gatherer bands, chatting about it the whole way.

It wasn't until ten thousand years ago that the next major development occurred. As the last ice age thawed out and global temperatures rose, humans began learning to grow crops and domesticate animals, in the so-called "Neolithlic revolution". In the same way that Hera was part of the Great Leap Forward, Ram was at ground zero for the Neolithic Revolution. It was agriculture that made complex civilizations possible, turning us from a species of tallying hunter-gatherers into mathematically-savvy urbanites.

Of course Egypt wasn't the only civilization in the ancient world. They popped up all across the globe, and Ram ended up having kindred spirits everywhere from the mountains of Peru to the Yellow River in China.

Traditionally historians identified four independent "cradles of civilization": Egypt, Mesopotamia in modern Iran, the Indus Valley in India, and the Yellow River in China. But this traditional catalogue is hopelessly Western-centric, and sorely in need of revision. It completely ignores, for example, the Norte Chico civilization in Peru, which probably had the highest population density in the ancient world, as well as the Mayans and Inca.
There is no mention of the Bantu people of Africa, who created the Great Zimbabwe city-state. This accounting also leaves out Papua New Guinea which, while it never built city-states, may have been the first place in the world to develop agriculture.

There's no correct litmus test for what constitutes a civilization. You could base your definition on agriculture, large cities, irrigation, writing systems or a host of other factors. Different civilizations mixed and matched all of these properties, making a good definition almost impossible to put your finger on. One is reminded of Supreme Court Justice Potter Stewart's definition of pornography: "I know it when I see it". For purposes of this book, however, we're looking at the history of mathematics, which means we need a written language and records of how they people with numbers. That rules out most civilizations, who either never had a writing system or whose writings haven't stood the test of time. Besides Egypt that leaves us with two other "cradles of mathematics" that we can study: Mesopotemia and China.

There are sixty seconds in a minute and sixty minutes in an hour. If you spin all the way around so that you are facing the same direction again we say that you've turned 360 degrees, which is equal to sixty times six. Given that our number system is generally based on ten, it seems a bit weird to measure time and rotation in groups of sixty. Where did that come from? This little bit of confusion comes courtesy of ancient Mesopotamia, where people developed the first documented system of mathematics beyond tally sticks, as well as what's usually considered the oldest full-fledged civilization. Our base-10 number system comes from India, by way of the Arab world, but the continued use of 60 is a holdover of Mesopotamian mathematics, which used a base-60 number system.

Mesopotamia is located in modern day Iraq and Iran, nestled between the Tigris and Euphrates rivers. These rivers carry water and nutrient-rich silt throughout the region, making it a prime area for agriculture. Mesopotamia is one arm of the the so-called "fertile crescent" - the other arm being the Nile River where Egypt later coalesced. Between 4 and 5 thousand B.C.E a collection of city-states sprouted up in the south of Mesopotamia, a region called Sumer. They invented the technology for making bronze, irrigated their fields, and constructed large brick temples called ziggurats. They also developed the first known writing system, which they wrote on clay tablets. Unlike flimsy Egyptian papyrus, clay tablets have the longevity of a fruitcake, so a lot of them are still around. The tablets that have been recovered show that the Sumerians also made great strides in medicine, philosophy and science.

As early as 8000 B.C.E. they were using clay tokens, marked with a symbol, to count sheep and other livestock. Each token meant one animal - a pretty straightforward system. For larger herds, there were other tokens which could denote ten or even a hundred animals, a lot like coin money today. By around 3500 B.C.E. physical tokens were replaced by writing the symbols directly into clay and baking it, which was the same way that they wrote their language. Tellingly though, they still used different tokens to represent sheep, goats and other livestock. Nobody was trying to represent "numbers" as things in themselves - this was all just a pragmatic way to keep track of specific goods.

The Sumerians didn't have just a single counting system. Besides using different symbols for different commodities, each city-state had its own variants. It took a long time for these systems to converge, across both regions and things to be counted. By around 2100 B.C.E though the city-state of Babylon had swallowed up the whole region into a single unified empire, and everybody converged on the so-called "sexagesimal", or base-60 number system. The sexagesimal system wrote numbers completely in the abstract, so that the Babylonians no longer had to count *things* like sheep; they could just count. Numbers were entities in their own right.

For the numbers 1-60, the sexagesimal system was similar to the token-based systems originally used for counting livestock. There was a symbol for 1, writing it twice meant 2, three times meant 3, etc. For numbers larger than nine there was an additional symbol for 10. A number like 32 would be written as three of the symbols for 10, plus two of the symbols for 1. This is the same method the Egyptians later used, just with different symbols, and it was the same way for any number less than 60.

However, the number system changed entirely for numbers from sixty onwards. Rather than a sign-value system like the Egyptians have, the Babylonians invented a system similar to what we use today. Sixty-three wasn't written as six symbols for ten and three symbols for one: it was written as the number one (a single 1 symbol) followed by the number three (three 1 symbols). This translated to "a single sixty, and three ones". It's the same way that in our numeral system "50" means "fives tens and no ones".
This is called a "place-value" system, because the meaning of a digit depends on where it is in the number. We have the ones place, tens place, hundreds place and so on. The Babylonians had

the ones place, the sixties place, and the three thousand six hundreds place.

The difference of course is that we just have ten little digits 0-9, whereas the Babylonians had an unwieldy fifty-nine. In our system the digits 0-9 are primitive symbols that we just memorize, but the fifty-nine Babylonian "digits" were themselves constructed from a simple sign-value system. It's kind of a weird hybrid, but once you accept those numbers as "digits", the overall logic becomes a sign-value system.

What I love about Sumerian numerals is that it showcases the evolution of writing systems. You can easily picture a society of farmers who get along just fine with simple tallying, exactly like Hera used. Eventually their flocks get big enough that, every now and then, tallying become unwieldy. Of course they don't want to throw out their perfectly good tally system just because a few jokers have massive flocks, so they jury-rig it by adding in a new symbol for ten. Things go fine from there, and as flocks continue to grow this new symbol becomes commonplace, until everybody is using it to keep track of animals. However, the biggest flocks again start pushing the system to the breaking point. Some people have hundreds of animals in their flocks, and who wants to juggle around twenty of the "10" symbols? So it's back to the drawing board again: how can we once again jury-rig the system to work for exceptional cases, but leave the smaller numbers the way they are? Eventually somebody thinks up the place-value system, and it ends up being good enough that it lasts for the remainder of their civilization.

The same pattern, incidentally, has persisted to the modern day. Normal Hindu-arabic numerals are good enough for most applications in daily, and even professional life. However, for a lot of scientific applications researchers are dealing with numbers so large or so miniscule that the number system becomes unwieldy. For example there are estimated to be about 1,000,000,000,000,000,000,000,000 stars in the universe. Forget arithmetic by hand - would you even trust yourself to enter that number correctly into a calculator? In what's termed "scientific notation" this number is written as 1.0×10^{24}, meaning a one with twenty-four zeros after it. Again in scientific notation, the radius of a hydrogen atom is about 5.29×10^{-11} meters. In certain areas of mathematics, where they study numbers that are just obscenely large, even scientific notation becomes impractical and they resort to other tools.

As impressive as their number system was for the time, there one major thing we have that the Babylonians didn't have: zero. The digit "0" functions as a place holder in our number system, so that 50 means "five tens *and no ones*". The 0's only purpose is to make it explicit that "5" is in the tens place. The Babylonian system didn't have any such place holder. It you wrote the digits five and four next to each other it wasn't clear what number you meant. It could be 5*60+ 4, but it could also be 5*(60*60)+4, or 5*(60*60) +4*60. It's like if we couldn't distinguish between 51, 501, and 510.

Despite this shortcoming, Babylonian numerals were the most powerful number system available at the time. They enabled the people of Mesopotamia to not only run their economy, but also develop sophisticated bronze-age technologies and even predict solstices and astronomical events. They recorded their mathematical knowledge on clay tablets, which have been able to endure thousands of years into the present day, as a lasting testament to their acumen. Hundreds of those tablets have been recovered, containing everything from multiplication tables to solving algebra equations, and even a highly accurate estimate of the square root of two.

When it comes to raw calculation power, the Oscar goes to ancient Chinese. By 1000 BCE they had already developed a full-fledged decimal-based numeral system, like we use today, which made rote calculations immensely easier. This enabled the development of a robust system of arithmetic, including algebra and even negative numbers. Armed with these tools, the Chinese performed detailed astronomical calculations, enabling them to predict eclipses, and also developed a powerful calendar system to aid in farming. The Zhou dynasty included mathematics in the "six arts" that were required for a well-rounded gentleman's education. The other five were music, calligraphy, Confucian rituals, archery and charioteering.

Unfortunately much of this development came to a dramatic halt in 212 BCE, under the command of the Emperor Qin Shi Huang. Qin Shi Huang, also known as the "First Emperor", was the guy who united all of China under a single ruler. Qin Shi Huang created a large bureaucracy to govern his sprawling empire, and appointments were based on merit rather than hereditary rights. He standardized currency, weights and measures, and created the framework of a unified China that has endured to the present day. No other ruler in history can make such a boast.

Sadly though, the order Qin Shi Huang brought came at a steep cost. He was a brutal ruler and a megalomaniac, practicing a style of government similar to modern totalitarianism. Thousands of lives were spent on his various buildings projects. Probably the most extravagant of these - and certainly the most self-centered - was an army of thousands of life-sized terracotta statues to guard his future tomb. But as over-the-top as his tomb was, it was only there as a backup plan. Qin Shi Huang wanted to live forever, ruling China in perpetuity, and he relentlessly sought out alchemists and sages in the hope of becoming immortal.

Among many grievances, Qin Shi Huang's most unpopular initiative may have been burning all books in China that were not sanctioned by his government. That meant all math books, all history books, all literature. If it didn't have his explicit approval, it had to go. Then, just to make sure everybody got the point, scholars who resisted were buried alive. Of course the decree was not universally obeyed, but it put an understandable damper on the mathematical ambitions of the next generation. Not to mention making it a lot harder on those of us today who are writing books that discuss the history of Chinese math...

Perhaps understandably, the Qin dynasty was not exactly popular. Qin Shi Huang died in 210 B.C.E. - ironically by mercury poisoning from one of his various immortality potions. Three years later - a mere 15 years since the empire was unified - Qin Shi Huang's son was assassinated during a popular uprising, and replaced by a former peasant named Liu Bang. Liu Bang founded the Han dynasty, which presided over a relative Golden Age for over four hundred years.

The great masterpiece of Chinese mathematics is Nine Chapters on the Mathematical Art. The earliest manuscript remaining is from early in the Han dynasty, though there is a chance it was originally written earlier and survived the book burning. Like the Moscow Papyrus from Egypt, it is a series of mathematical problems with worked solutions. However, while the Moscow Papyrus had only 25 problems, the Nine Chapters contained a staggering 246 of them, covering everything from agriculture, to land surveying, to the properties of triangles. Greek mathematics, which I'll discuss in the next chapter, was developing around this same time, synthesizing results from Babylon and Egypt into fancy theoretical frameworks. In contrast, the mathematics of China was almost entirely practical. It focused on general ways to solve problems, and in the process developed a radically different approach. The Chinese failed to discover many of the "elegant" results that the Greeks were very taken by. On the other hand, they developed systematic ways of solving problems which were not fleshed out in the western world until a millennium later.

While farming and commerce were almost certainly the main drivers for early math, they were definitely not the only ones. I would be remiss if I didn't mention the roles played by astronomy and religion.

Astronomy was in a lot of ways the first of the mathematical sciences. From Stonehenge to the Mayan calendars, ancient people were fascinated by the stars and planets. This was one of their biggest motivators both in mathematics and technology in general, and perhaps the main thing that forced them to push well beyond basic arithmetic to something as fancy as a place-value number system. It's a lot more complicated to calculate when Mars and Jupiter will align than it is to balance a checkbook.

I'm talking about all of this in regards to ancient mathematicians, but really the trend has continued to this day. In the same way that star charts may have motivated the Babylonians' number system, they also pushed forward the study of geometry by the Greeks millennia later. In the 1700s, Newton invented calculus to describe the force of gravity, and how it could be used to predict the shape of planetary orbits. Now in the modern era, astronomers are developing new computer tricks to analyze their datasets. And this is to say nothing of the technical aspects of telescopes!
Accounting is mathematics' real sister discipline, but astronomy is at least a second cousin, who constantly goaded it on.

Religion doesn't have much to do with math these days, but it definitely did in the ancient world. There were many indirect connections, like organizing tithes and using astronomy to calculate the dates or holy days. In some cases it went beyond that as well. My favorite example is ancient India, where different gods required altars of different shapes. That's all well and good, until you realize that the altars had to be exactly the same size lest some diety take offense. It's easy to make a square and a triangular altar about the same size, but how can you make them so close that even a god can't tell the difference?

There are two things I want to call out about early mathematics, that apply to every place it arose. First, it was extremely problem-oriented, with all of the existing manuscripts based on worked examples instead of overarching theories. Secondly, the written number system was of paramount importance.

To the modern eye, the ancient world's focus on mathematical problems, rather than abstract mathematical theory, is often seen as a deficit. So much so in fact that everything in this chapter is sometimes called "pre-modern" mathematics, since it was done without proofs and logical deduction. But that label is not really justified, especially given how far China got without proofs. I'll talk more about rigorous math in the next chapter, but it's not fundamentally an improvement on the more empirical math in this chapter - it's just a different attitude. The phenomenal success of Chinese math is evidence enough of that.

It's easy to understand the importance of picking a good number system. If you're doing all of your calculations by hand then it's critical that the numbers are easy to work with. In this regard, the Hindu-Arabic numerals we use today are a virtual miracle. They are concise, with very little writing required to record a number. They are precise, with none of that bullshit about missing zeros that you had in Babylon. The symbols themselves are easy to write, unlike Egyptian hieroglyphics. And above all, having a place-value system made it easy to do calculations. Long division may be an unpleasant prospect no matter what number system you're using, but imagine doing it with Babylonian numerals! We are now in possession of the greatest number system in human history, but previous societies developed their own clever, ad-hoc methods to deal with numbers, as the problems they solved got more and more sophisticated.

One side effect of a good number system is that math becomes abstract. You don't have to add sheep to sheep, or multiply barrels of wheat per land parcel by the number of parcels. You can just do arithmetic on *numbers*, and then apply the same techniques to problems in any area. It's tremendously economical to use the same arithmetic tables for architecture that you use for paying taxes. Linguistically, numbers graduated from being adjectives that describe a group into nouns in their own right.

I'd like to close this chapter by circling back to the thesis of this book, that math is really just a branch of language. I may have convinced you that tallying is language, but isn't calculating square roots a stretch?

Well in a sense yes. This chapter is where mathematics and natural language part ways forever, going down separate evolutionary paths. Natural language proceeds much as it always has, except that vocabulary expands and contracts to keep check with the times. Mathematics, on the other hand, veers off at a tangent, away from our intuitive scope and toward more rigor and abstraction, getting further and further removed from its earthy origins.

These superficial differences don't mean a fundamental change. Math is *more* abstract than natural language, but natural language can also deal in abstractions. Take the noun "red", for example. "Red" isn't a substance in the world like water, that you can go out and touch. It's just a property that some things in the world have, and we encounter red objects so frequently that we created a whole new noun to describe the color. In the same way, many objects in the world come in pairs, and the word "two" is just the linguistic embodiment of that fact. The sentence "2+3=5" is no more or less abstract than "red and yellow make orange".

Underneath all the layers of abstraction math is still talking about the real world. The bizarre evolution of math doesn't absolve us from having to understand it. It just means we have to work harder. The ancients understood this, and that's why they always taught math in the context of its applications. Math is unnatural enough as it is; there's no need to exacerbate the problem by making it unrelatable as well. Numbers are still there to describe reality, and math problems give us facts about the world around us. Modern educators could learn a thing or two from Ram.

4 PROOF AND PREJUDICE

My first quarter of college got off to an ambitious start. I was fresh out of high school and dead-set on becoming a scientist. Sadly this meant that I took the fewest humanities classes humanly possible (ironically, the single coolest class I took in college ended up being on gender studies), but it also meant I boned up on cool math and science. I enrolled in honors physics, where we studied how pendulums swing, balls roll down hills and planets orbit around stars. This is the class where I met the core social group for my years of college, and it was also here that I learned the intuitive way that physicists think about math. They spend a lot of time chugging through equations (as I know from many sleepless nights...), but they come up with those equations in the first place through sketches on a chalkboard, pictures in their heads, or imagining how a "thought experiment" would turn out.
The math itself, though often very advanced, plays second fiddle to the underlying physics. The real emphasis in the curriculum is on understanding the physics so well that you can describe it using equations.

At the same time I enrolled in a course that you might think would be very similar, but that ended up being radically different: honors multi-variable calculus. According to the course description, it was just supposed to cover the math that I would be using in my physics class. But in the first lecture the professor explained that no, this would be unlike any other class I had ever taken. It wouldn't teach us a lot of new material, but it would teach us "mathematical maturity" - the peculiar, hyper-rigorous way that professional mathematicians approach their subject in this day and age. I started to get nervous as he went on and on about how difficult this class would be, how we would spend long hours proving theorems and how most of us would eventually quit the class in defeat. How many credit hours was I signed up for again? And how many of those were already taken up with masochism-fest physics? I began to sink lower in my seat. A couple lectures later the professor showed us a long, intricate proof that $0+0=0$, and I was out the door.

I've often wondered what my career would have been like if I hadn't chickened out. My fear is that I would have ended up drinking the kool-aid of modern mathematicians, internalizing the dogma that "real" mathematics is all about obsessive, nit-pickingly rigorous proofs. I took other classes later, and eventually developed this storied "mathematical maturity", only to find that it's just one tool in a large palette of ways to look at math.

I'll talk in later chapters about how mathematicians found themselves in the seemingly absurd situation of trying to prove that $0+0=0$. But they took the first steps down that road in ancient Greece. Greek philosophers were very taken with the idea of human reason as a tool for making sense of the world. They tried to organize prior knowledge into overarching intellectual frameworks, using deductive reasoning to show things from first principles rather than just sticking with empirical rules of thumb.
They tried to systematize everything from politics to science into these grand frameworks, but nowhere was their influence more profound than in math. The introduction of proofs changed the entire course of mathematics, at least in the Western world, so much that some people dismiss everything that came before as "pre-modern".

This chapter of the book will include three short proofs, interspersed with the discussions of people, events and ideas. I only put them in to illustrate what goes into a proof, which many of my readers won't have seen since high school geometry. You don't need proofs to understand why math is a language, which is the key theme of this book. However, you do need to know what goes into a proof in order to understand the last two thousand years of math's evolution, so I want to make sure you get a taste of them.

I have been warned that when you are writing a book for a general audience - even a book about math - every equation or proof you include will cut your sales in half. Instead, authors are supposed to dance around the subject, limiting themselves to vague generalizations and pretty pictures. Heaven forbid there be equations in a math book! But I wouldn't want to write such a book; it would be like writing about how the US Congress operates without ever daring to explain what a bill is! And frankly, I think my audience is better than that. A popular book should never get into needless technical details, but many authors cross the line from simplifying a subject into misrepresenting it. So yes, I will show you what an honest-to-goodness proof looks like. And in the process, hopefully I'll show you that they're not as scary as most publishers think.

If you had to pick one person as the father of Western thought, there's a strong argument to be made for Thales of Miletus, a Greek philosopher who lived around the year 600 BCE. Thales isn't the most famous philosopher from ancient Greece. He lived at a time when philosophy was just gathering steam, and he didn't leave behind any big written treatises. In school today you're more likely to hear about Socrates or Aristotle, both of whom came centuries later. They were the ones who really fleshed out Greek philosophy (or at least, one particular school of it), but it was Thales who blazed the trail.

He was the first person known to have used deductive reasoning as a tool in philosophy, including writing the first formal mathematical proof. Thales also applied geometry to many practical problems in navigation, astronomy and architecture (including, famously, measuring the height of the Egyptian pyramids based on the length of their shadows). It even looks like he was an accidental pioneer in economics, inventing the idea of options as financial instruments for trading. However, Thales' most important legacy was not in mathematics or finance, or even any specific disciple. It was the approach he took to all problems.
 Thales looked for natural, rational explanations for how the world behaved, instead of just writing things off as a mystery or the will of the gods. This idea was taken up with zeal by later thinkers, and formed the core concept in the scientific revolution.

We don't know much for certain about Thales' life, since he lived so long ago. From what we do know it looks like he came from a wealthy family, because he was able to travel widely while growing up. This seems to have included Athens, the cultural and intellectual hub of Greece. While there Thales met with Solon, the legendary statesman who established Athenian democracy. Thales even ventured as far as Egypt, absorbing all the knowledge he could along the way.

As an adult Thales was less of a "philosopher" in the modern sense, and more of a politician who did philosophy on the side. Specifically he was a close advisor to king Croesus of Lydia, especially in military matters. Defense was a big deal at the time, since Greece was divided into a number of city-states who were frequently at war with one another. This was bad enough, but at the same time the Persian empire was growing in the East, threatening to engulf the whole region if they couldn't stop squabbling.

The details are murky, but it looks like in 585 B.C.E. Thales managed to correctly predict an eclipse of the sun. The fulfillment of this prediction stopped a battle between Lydia and the neighboring city of Medes dead in its tracks. Apparently both sides took it as some kind of omen. Instead of fighting each other the two cities joined forces against an invading army from Persia, thwarting what might otherwise have become the Persian conquest of Greece. Thank you astronomy! The Persians would try and fail again several centuries later at the Battle of Thermopylae, which was recently dramatized in the movie 300.

A lot of people think that Thales only took a serious interest in philosophy after he retired from full-time politics. I can't help but draw a comparison to Benjamin Franklin, who spent most of his career as a newspaper publisher, doing some random science and inventing on the side. After retiring from publishing Franklin got into science full-time, and made major discoveries about electricity that launched him to international stardom, like a Stephen Hawking of yore. Later, it was Franklin's scientific renown that gave him the clout to negotiate with France during the American Revolution.

In exploring the natural world, Thales was fascinated by the the fact that the same materials, under different conditions, could take on a range of properties. Metals turn liquid when heated, and back to solids when they cool down. Dirt gets absorbed into plants, which in turn can become part of animals or decay back into dirt. The most transformable material of them all, at least that Thales would have been familiar with, is water. It is the only substance that can be frozen into a solid, melted into a liquid or boiled into a vapor using simple technologies. If you don't have a microscope, snow and ice look like entirely different substances. In fact, snow is arguably a lot more like dirt than it is like ice, and ice in turn looks an awful lot like crystals of quartz. Even the most outlandish matter in the world - living creatures - is composed mostly of water.

Thales hypothesized that water was, in fact, the basis of all matter. Quartz was just another type of ice, one that for some reason didn't melt at room temperature. Dirt was a brown, heat-resistant form of snow. Plants and animals were just forms of water too. The world itself was one big ocean, out of which the continents had crystallized, and they float in it like ice cubes. The things that we call earthquakes are really just waves in the gigantic ocean.

This all might look pretty absurd to modern sensibilities, but given the data available at the time it was very plausible. Even though Thales' choice of water as the master substance was incorrect, he was definitely on to something. Over two thousand years later we now understand that all normal matter is just combinations of protons, neutrons and electrons. But I digress.

While Thales was traveling in Egypt he became an avid student of their geometry. He made extensive use of triangles, applying them to practical problems such as calculating the distance to a ship at sea, and even using an early form of trigonometry.

This work led him to compose the first known deductive math proof in history, demonstrating what's known as "Thales' theorem". Thales' theorem, and its proof, are very simple as proofs go. Like most proofs in geometry though, they're a lot easier to follow if you draw out a picture and keep referring back to it. In this case, the theorem is illustrated in this figure:

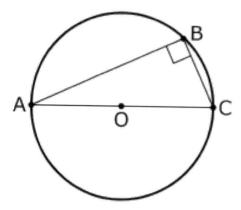

Thales showed that if you cut a circle in half, and then connect the endpoints (A and C) to any other point on the circle (B), you form a right triangle. That is, Thales' Theorem says that the triangle's corner at point B is ninety degrees in the picture I've drawn, no matter where on the circle B is. Looking at the picture this conclusion is certainly plausible, but at least to me it's not *obvious*. I certainly wouldn't bet my life on it just because the picture looks right. The Egyptians had probably discovered Thales' Theorem by trial and error long before Thales' time. They used it in daily calculations with no ill effect for years before Thales came along and hammered out a rigorous proof.

To prove the theorem deductively, Thales broke the triangle in two by drawing a line from B to the middle of the circle, splitting it into two smaller triangles:

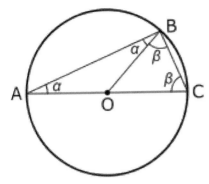

The way I've drawn them here, triangle OAB is fairly pointy while OBC is more short and squat. This is just an illustration though; I could have drawn it another way by putting B somewhere else on the circle. But the key thing to notice about these triangles, which would have been true no matter where I put B, is that they are equilateral. The length of OA is the same as the length of OB, and similarly OB is the same length as OC.

Equilateral triangles have the really useful property that two of their angles are the same size. Let's call them and in this case, which I've drawn in the picture. In this case happens to be smaller than , but again that's just because of how I've drawn the picture. What's always the case though is that the angle at B is + , while the angle at A is and the angle at C is . So the angle at B is the sum of the angles at A and C.

There is one more well-known fact that we need in order to complete the proof: the sum of all angles in a triangle is 180 degrees. We showed in the previous paragraph that B's angle is the sum of the other two, so it will be half of 180 degrees, i.e. 90 degrees. That means it's a right angle, and overall we have a right triangle!

If you ask a professional mathematician today they're likely to tell you that deductive reasoning is the defining characteristic of mathematics. Now first off that's an absurd definition, since it glosses over the first fifty thousand years of the subject, including brushing off the tremendous accomplishments of the ancient Chinese and Egyptians. More importantly though, logical nitpicking is not specific to mathematics. Deductive reasoning is actually a generic property of languages.

Take legalese, for example. It can be dense, counterintuitive stuff, requiring the same double-checking and bean counting as a mathematical proof. Some lawyers I know complain that their discipline is too abstract, a head-in-the-clouds game of logic and definitions that can leave them hungry for something more real.

Deduction is especially important in areas like patent law. It's often the case that everybody on both sides agrees on the facts of the situation. Who said or did what, what money changed hands, what order events happened in, and so on. However, whether those events actually constitute a crime is not immediately obvious, given the intricacies of individual patents and the laws governing them. It's pretty rare for large corporations to sue each other over intellectual property. What happens instead is that the lawyers on each side construct deductive "proofs", carefully ferreting out which laws apply to the situation and what they prescribe, and determine how a hypothetical trial would turn out. There's no rhetorical polishing or emotional appeals to a jury. Just a clear, irrefutable chain of logic that shows who is on the right side of the law. The losing party then cuts a check, and everybody avoids the hassle of an actual lawsuit.

Of course, math takes deductive reasoning to an extreme that goes well beyond even the densest legalese. The whole syntax of math has been re-worked to be more precise and logically consistent than is usually possible with a language like English. It is so over-the-top that, in practice, people often treat math as just a game of manipulating symbols, rather than descriptions of the world. And while this is an easy way to cop out, it is dangerous. The precision and certitude of math is illusory. Insofar as we're talking about the real world, any language gives only a partial picture. If we decouple that language from actual understanding, we have no safety net when the language fails.

There are other reasons to do proofs, besides just making sure our theorems are correct. This is a part of the book where I get to geek out a little bit. I think this stuff is awesome enough that I've made a career out of it, and I hope to show you why I find it so cool.

Personally, I love reading a really clever or simple proof. I like to dwell on the good ones, not just making sure that the logic checks out, but really chewing on them until I internalize how the author was thinking about it. What insight did they have into this problem that inspired them to formulate the proof the way they did? For example, what possessed Thales to split up that right triangle the way he did? This exercise builds my intuition for the subject, forcing me to explore its depths and think about it in new ways. The very best proofs are hardly proofs at all. They are a new way of framing or rephrasing the problem, which casts light on the issue and makes the theorem seem obvious.

Some people would argue that rigorous proofs dehumanize an area like geometry. It goes from being a beautiful game of pictures and shapes to a sterile chore of checking definitions. Does that step work out logically? Did I define this term?

Frankly, there's a lot of truth in this criticism. Deduction often sacrifices beauty for precision, and some proofs don't really contain the insightfulness or intuition that I like so much. If you're like me, the proof then shifts from being "elegant" to being a fun little logic game, like Tetris or minesweeper. Working through a tricky step in a proof can give the same intellectual rush as finishing a sudoku puzzle. Now I'll admit that some proofs are just miserable slogs to read through, but in practice that tends to be the minority. In fact, many mathematicians have gotten famous not by proving new theorems, but by coming up with beautiful proofs to replace the old ugly ones.

Of course, my rosy sentiments about math are not universally held. Even though I love reading clever proofs, for other people all of mathematics is just pulling teeth. That's a tragedy, and I really hope this book can help change a few peoples' minds. You might not be a musician, but you can still sing in the shower and have a favorite band. You might not be an athlete, but you can still enjoy a brisk walk and cheer at a football game. Math is the same way. When taught properly it has the same kind of visceral appeal as music or athleticism. Its logic and clarity resonate with a deep part of the human mind, and any person can learn to appreciate that.

At the end of the day though, these sentiments are neither right nor wrong. What's right or wrong are the formulas and the equations. Large numbers and detailed geometry are beyond anybody's innate intuitive scope, and we need a mental crutch to deal with them reliably. Love it or hate it, being formal is the price you pay for being correct.

At some point in his career, Thales become the mentor of another aspiring philosopher named Pythagoras, who became his most famous student and one of the greatest mathematicians of the ancient world.

Now Thales was a more-or-less normal, practical guy. His mathematical work was pretty theoretical, but it was motivated by, and applied to, real-world problems. His scientific studies weren't necessarily practical, was at least they were down to earth. He was an active member of his community, even winning awards in local athletic competitions. And after a long day at work advising the king, he went home at night to his wife and adopted son. Really a role model all around.

Pythagoras, on the other hand, was an honest-to-goodness cult leader. He invented a religion that hybridized Greek mysticism with a reverence for numbers and geometry, attracted a large number of followers, and retreated with them to the Italian city of Croton. Thankfully the Pythagoreans didn't just spend all their time prostrating themselves before the number 42. They were insular and secretive, but they also doubled as an order of teachers and scholars, kind of like Jesuit priests do today. They produced many important mathematical results (almost all of which they attributed to Pythagoras personally, regardless of his actual involvement) - including a proof of the Pythagorean Theorem - and trained many of the leading minds of the time.

Pythagoras, as far as we know, never wrote anything down, so what we know about him and his cult comes from contemporaries and ancient historians. Even his contemporaries had a hard time getting good information, since the Pythagoreans were notoriously tight-lipped about their specific beliefs and rituals. We do know that Pythagoras saw in numbers a purity which transcended the crude physical world around us. He believed that numbers were not abstractions for describing the physical world, but rather a mystical foundation on which the physical world was based. The highest pursuit of the human mind was not in the living of our daily lives, but in stepping back to contemplate the transcendent nature of reality.

The Pythagoreans believed in reincarnation. Centuries later Julius Caesar would speculate that the druids of England, who also believed in reincarnation, picked it up from them. They taught that the cycle of death and rebirth could be broken by purifying the soul and focusing on transcendent ideas, eschewing the distractions of daily life. The end goal is for the soul to leave the death/birth cycle and take on a contemplative existence, meditating on the mathematical patterns of reality. Pythagoras preached a simple lifestyle, free of physical indulgences that could distract the mind from its transcendent pursuits.

Personally, I can't help but blame Pythagoras a little bit for some of the negative stereotypes that have plagued mathematicians. I mean, what does it say about you if one of the luminaries in your field literally founded a cult that worshiped numbers? Thales on the other hand was a respected leader in his community, a family man, an amateur athlete, and above all he was non-crazy. Couldn't we be famous for him instead? But the history books have been written, and for better or worse, Pythagoras deserves his place in them.

Pythagoras' cult lives on today in many college math departments.

Most mathematicians are what's called "Platonists". This means they think that numbers, triangles, and other mathematical objects aren't just human creations. They "exist" in some absolute, transcendent sense. For a Platonist, doing math research isn't just inventing new techniques or refining definitions. It's discovering new facts, about objects that already existed. It's less like engineering or invention, and more like exploring a new continent. This is because numbers are as real as the world around us... and possibly even moreso.

When I come across a Platonist in conversation, and they start talking about how the Pythagorean Theorem as some sort of transcendent truth, I like to ask them a simple question. If the Pythagorean Theorem is such a pristinely true statement, then what exactly is it a statement *about*? No triangle in the real world will ever have exactly a 90 degree angle, or a side that is perfectly straight. So what, pray tell, are these mathematical objects that are supposedly so real?

When you press Platonists on this point, they tend to start muttering and trail off. Eventually they retreat into saying that math is just a meaningless game of symbols. They finally concede that the Pythagorean Theorem isn't really a statement about anything; it's just that last line in logically valid math proof. However, that's really not satisfactory either. There may not be any perfect triangles in the world, but there are some damn good ones, and the Pythagorean Theorem describes them awfully well. It is a sort of hypothetical idealization, true in the limit of a perfect right triangle even if it's only approximated in the real world. On some level you can view math as a game of symbols, but there's gotta be more to it than that.

The Platonists in math take their name from the Greek philosopher Plato, who tackled this question head-on and took it to a logical extreme. For Plato, a "perfect right triangle" isn't just hypothetical. It is a very real entity, just not a physical one. It is what's called a Form, the transcendent embodiment of what it means to be a right triangle. For every mathematical entity there is a Form, and the Forms reside in a sort of metaphysical plane of reality called the Realm of Pure Forms. If this is starting to sound a little cooky, and like Plato might have been smoking something, then I agree with you wholeheartedly. But bear with me.

The Pythagorean Theorem is an absolute, rock-solid fact about the Form of Right Triangles. Timeless and exact, no ifs ands or buts. The theorem is useful in daily life because the Form of Triangles is crudely approximated by those things we call triangles in the real world. Or excuse me: those things we call triangles in the *physical* world. The Realm of Forms is the "real" world in Plato's mind, and the world around us is just a cheap knock-off. Even the fingers on your hand only get their "fiveness" via a metaphysical link with the Form of Five.

Plato's Forms aren't limited to mathematical entities. There is also a Form of Chairs, and the office chair I'm sitting in right now has a metaphysical link to it. My office chair only gets its chair-ness from that link. There are Forms for "nationalism", "music" and "tree". Above all there is the Form of Goodness, and studying it is the central task of philosophy.

Plato said the things in the physical world are like shadows cast by the Forms, in the same way that physical objects can cast a shadow on the wall. A shadow gets its size and shape of a from the thing that casts it, but the shadow itself is just a basically a non-entity.
To show how this worked, Plato invented the Allegory of the Cave, one of the most famous stories in philosophy. Imagine a prisoner in a cave, chained from birth so that they can only look directly at a wall. There is a fire burning behind him, and it casts shadows on the wall of whatever else is in the room. The prisoner, having spent his entire life looking only at shadows, will think that these shadows are actually the real objects of the world.

The prisoner of course will be oblivious to his miserable station in life. If empty shadows are all he's ever known, that's all they'll need to be content. Eventually he will learn to recognize that there are different kinds of shadows. Big blobs, small blobs, shadows that branch out like trees, and so on. He may even start thinking he's pretty clever, congratulating himself on developing a fine taxonomy of shadows and detailed knowledge of their workings.

Plato saw most people of the world prisoners like this, staring at shadows on a wall and thinking that's all there is to the world. Philosophers on the other hand, who spent their time studying the Forms, were like prisoners who had escaped the cave, seen the world as it truly is, and were now trying in vain to bring enlightenment back to an ignorant populace (it makes you wonder what Plato's social life was like).

In what amounts to a fleshed-out version of Pythagoras' cosmology, Plato taught that our souls are reincarnated, returning to the Realm of Forms between lives. This is the reason I can say something in this world is "circular", even if it's not perfectly so. It reminds me of the Form of Circularity. My soul perceives the metaphysical link between a circle drawn in the dust and the Form of Circularity. I can say that something is more or less round only because my soul has a basis for comparison. In between lives, my soul has actually seen a perfectly round circle, and on some subconscious level I use that as the gold standard.

Plato's ideas about the Realm of Forms are really an off-the-wall theory of language. The existence of Forms gives a concrete meaning to words we use, because every word now corresponds to a very real (even if non-physical) object. When language doesn't describe the world perfectly, it's not the fault of language. It's the fault of the world, because it only crudely approximates the Forms.

On the off chance that it hasn't been clear so far, I think the Realm of Forms is baloney. To my mind Plato was somewhere between Thales and Pythagoras. He was a creative genius, contributing brilliant new ideas across all areas of philosophy. At the same time though, his grasp of the big picture is more than a little dubious, even if he didn't go totally off the deep end and found a number-worshipping cult. He may have been a genius, but everything he says should be taken with a hefty grain of salt.

Words like "Rome" correspond to actual things, but I'll never understand why a bright guy like Plato couldn't accept that other words just don't. He put language such on a pedestal that he made up a transcendent la-la-land to embody it, rather than simply admit that language is imperfect. I can't help but snicker at the idea that our paleolithic ancestors, in between gathering berries and running from cave lions, stumbled upon a portal to a higher plane of reality when they invented language.

But there is one place where Plato deserves a serious tip of the hat. He understood that there is no fundamental difference between mathematics and the rest of language. If mathematical idealizations like "circle" and "five" have their place in the Realm of Forms, then "tree" and "indigestion" had better be there too. He was perhaps the first person in history to realize that the question "what is math?" reduces to the question "what is language?".
There is a fine line between a genius and a crackpot.

Plato's ideas never really caught on in natural language. Nobody to my knowledge believes that there is a metaphysical "Form of Reuben Sandwich" (unless they've been to Katz's Deli in Manhattan, of course). In math though it has become very popular, and a lot of modern mathematicians fancy themselves to be charting out new territory in the Realm of Forms. Many people a lot smarter than myself buy into this fantasy, but I respectfully think that they should have their heads examined.

The most successful textbook of all time - unless you count religious texts - is a geometry book called The Elements. If you had taken a geometry class anywhere in the Western world up through the 1800s, Elements would probably have been the textbook you used. Nowadays people don't use Elements itself much, but almost every introductory geometry book is just a re-packaging of the same material.

Elements was written by the Greek mathematician Euclid, who was active roughly around 250 B.C.E - some 320 years after Thales of Miletus pioneered deductive proofs. Euclid took this approach to its logical conclusion, stripping out all the vestiges of earlier empirical mathematics and re-imagining the discipline from the ground up. Gone was the cookbook mathematics of the Egyptians and Babylonians, where you just plugged new numbers into old formulas that had worked in the past. Real-world applications were out the window too. Math was still useful of course, but the geometric abstractions were not to be tainted by the physical world.

Euclid realized the key problem with logical deduction: it needs some first principles to start with. Every step in a deductive argument is valid for some reason, and conceivably somebody could poke holes in your argument by demanding a whole new proof that each step is justified, ad infinitum. So Euclid began his book by stating a handful of clear, basic assumptions called "axioms". The axioms were chosen to be as glaringly self-evident as possible, but after that point they were taken as gospel for the rest of the book. Every step of every proof was justified either by an axiom, or by a theorem that had already been proven using the axioms. Working forward in this way, step by logical step, Euclid re-created the whole of geometry as it existed at the time.

The irony is that Euclid may not have actually been that good a mathematician. It's not even clear that he discovered anything on his own. Mostly he was just re-packaging the findings of other men, like Thales and Pythagoras. His genius wasn't in doing math, but in organizing and teaching it.

And organize math he did. It's impossible to overstate the impact Euclid had. His precedent of axioms, careful deduction and massive abstraction set the tone for the next two millennia, continuing up through the present day. No book about math would be complete without highlighting this so-called "formal" or "rigorous" style of math.

At the same time though, it's important to remember that Euclidean formalism is not a natural way to look at mathematics. Deduction entered math late in the game because, for the most part, our brains don't work that way. Euclid did his best to sever the umbilical cord that tied math to the physical world and our intuitions about it. So learn about proofs and deduction, by all means! It's fascinating stuff, and even more important now than it was two thousand years ago. Just bear in mind that formal math is an acquired taste. It's a distorted, unnatural way to look at the subject, that went off on an evolutionary tangent even by the standards of mathematics. Trying to understand the nature of math by reading Elements is like trying to learn about mammals by studying the platypus!

Checking a proof for correctness is a science, but picking good axioms is an art. They need to be accurate, and to paint a complete picture of whatever it is you're studying. They also need to be clear and obvious, so that nobody will question them. As a practical, technical matter you want axioms that are easy to fit into a proof. There is no absolute right or wrong when picking these things, but there are a lot of judgment calls.

I'd like to bring up the analogy between math and the law again, because picking your axioms is very similar to writing laws. Euclid's axioms are one man's attempt to codify his intuitions about physical space; he took his understanding of how geometry works and tried to condense it down into just a few self-evident principles. In the same way, a body of law is a society's attempt to codify its sense of right and wrong. It's an interesting game to distill something that we understand intuitively down into a few precise statements. Sometimes the process reveals inconsistencies in our intuitions, which need to be resolved on the fly. We also usually find that no codification can capture all of our instincts perfectly. Everybody knows, for example, about the prevalence of loopholes in the legal system, and the futility of trying to close them all. Whenever you translate human intuitions into precise language something gets lost.

Nevertheless, the laws we legislate or axioms we choose are far from arbitrary. They are trying to codify very real physical phenomena, cognitive mindsets, and moral urges. Everybody agrees that murdering babies is wrong, and any set of laws should reflect that. There may be room for debate about the definition of "murder" versus "manslaughter", but it's usually very clear if we wrote our laws wrong and need to go back to the drawing board. In the same way, humans have very real intuitions about how numbers work and how space is structured. These are based on deep primal instincts like the number sense, and on our experiences living in the world. Axioms and laws function as first principles for deductive reasoning, a solid foundation on which other knowledge is built. But this foundation is itself laid on the not-so-rigorous bedrock of human intuitions.

For that reason, strictly speaking, axioms and laws are both aiming at moving targets. Societal sentiments change, and we modify our laws to capture that evolution. In math we sometimes find that one set of axioms isn't nearly as useful as we thought it was, and changes get made. Suppose, for example, that we found out mathematicians have been overlooking an error for two millennia, and Euclid's axioms don't actually support the Pythagorean Theorem. In that case we wouldn't abandon the Pythagorean Theorem. Instead we would go hunting for a new set of axioms, because the fact is our bridges aren't collapsing.

One of my professors in college used to complain that when you read a math book or research paper it starts with premises, builds up a clear argument and culminates in an elegant conclusion. Organized and tidy, it has everything except for a bow on the top. However, the actual process of writing that paper was completely the reverse. First the researcher figured out what she wanted to prove, based on her intuitions for the subject, a picture she scrawled on a napkin, or something along those lines. Then she worked backward, massaging her premises until she could prove the theorem she wanted. This is the way mathematics happens in the real world. After all, even Euclid was really just giving a fancier foundation for discoveries that had already been made;

there was surely a lot of trial-and-error in writing Elements and picking his own axioms. The discoveries are made through intuition or empirically, and rigor is introduced later to hammer out the details. But you'd never know it reading a math book.

Ok, so what were the axioms Euclid settled on? I'm not going to go very far down that road, since this book is about mathematics in general rather than geometry in particular. You need to know that Euclid's axioms exists, but it's not important to know them exactly (confession: I often need to refresh my own memory of what they are). However, they're very straightforward and illuminating, so I'll give you the rundown.

Euclid broke his axioms down into five "common notions", which applied to many areas of math, and five "postulates", which were specific to geometry.

The common notions were rules for when two shapes (line segments, 2-dimensional figures, etc) have the same or different sizes. Imagine that we have shapes A, B, C and D, whose sizes are a, b, c and d respectively. Translated from the original Greek and expressed in plain English, the common notions said
1. If $a=b$ and $b=c$, then $a=c$
2. If $a=b$ and $c=d$, then $a+c=b+d$
3. If $a=b$ and $c=d$, then $a-c=b-d$
4. If A and B are exactly on top of each other, then $a=b$
5. If A is a part of B, then $a < b$
This may seem like comically obvious stuff, but remember: that's the goal in choosing axioms!

The five postulates, which talk about geometry in particular rather than math in general, are a little bit more complicated. I'll say what they are in a minute, but first let me explain where Euclid was coming from. His goal wasn't just to prove theorems about geometrical shapes in general, but to actually draw the shapes at the same time. For that reason it's called "constructive" geometry. Most of his proofs are actually recipes for how to draw shapes, and along the way he proves that those shapes have right angles, parallel lines, or other properties. In order to draw the shapes Euclid imagined himself having two tools: a perfect straightedge that he could use for drawing lines between points, or for lengthening line segments out into longer lines, and a compass, like you might use on a map for drawing circles. Those tools are the only ones available: straight lines and circles. Note that there are no rulers, so Euclid isn't allowed to compare distances explicitly. There are also no tools for measuring angles. If you want to draw two line segments that are the same length, or two angles that are equally pointy, you'll have to figure out a way to do it with just the straightedge and compass.

With that prelude, the first four of Euclid's five postulates are
1. For any two points you can draw a straight line between them
2. For any straight line segment, you can extend it out into an infinite line
3. For any point in the plane and any radius, you can draw a circle that is centered at that point and has that radius
4. All right angles are the same size.
The first three postulates are the basic ways that we can construct shapes with a compass and straightedge, and the fourth postulate is there for reasoning about the shapes once they're drawn.

So those are the first four postulates. Euclid's fifth and last postulate, also known as the parallel postulate, is the odd one out. It's a little more complicated to state, and Euclid always felt like it was less obvious than the other four. Rather than state the parallel postulate as Euclid said it, I'll just tell you an axiom that is equivalent to it called Playfair's Axiom:
5. For a given straight line L and a point P that is not on L, there is exactly one line that passes through P and is parallel to L, as shown in this picture:

What is Math?

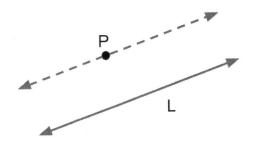

The parallel postulate is not wrong. In fact it's obvious - just not as glaringly so as the other four. But because he felt like it was more dubious, Euclid avoided the parallel postulate in his proofs whenever possible, so a large fraction of Elements doesn't actually depend on it. I'll come back to the parallel postulate a bit later, but for now take it as a given.

To give you a taste for Euclid's style of argument, let us prove his first theorem. This is the last proof in the chapter, I promise! The proof is shown in this picture:

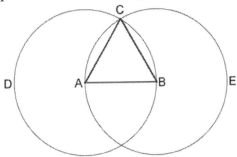

[Credit: http://www.storyofmathematics.com/hellenistic_euclid.html]

Theorem:
    If we have a line segment AB, we can draw an equilateral triangle with AB as one of its sides.

Proof
1) By postulate 3, we can draw circles around A and B, both of which have AB as the length of their radius. The circles are labeled D and E in the picture.
2) Pick one of the points where these circles intersect and call it C

3) By postulate 1, we can draw line segments AC and BC
4) Since points B and C are both on the circle around A, we know that segments AB = AC
5) Since points A and C are both on the circle around B, we know that segments AB = BC
6) Because AB=AC and AB=BC, we know from common notion 1 that AC=BC
7) Because AB=AC, AB=BC and now AC=BC, we know that all sides are equal to each other. This is the definition of the triangle ABC being equilateral.

Euclid filled thirteen books with arguments like these, constructing from the ground up all of the mathematics that was known at the time.

As a historical note, people tried for centuries to prove the parallel postulate from the other four. Doing that would tie up the final loose end, and put all of geometry on a *really* firm foundation, right? Euclid was the first person who tried to prove the parallel postulate, but he ultimately failed. So on down the centuries, through the middle ages, and even through the scientific revolution, countless people flung themselves against this wall. Many "proofs" of the postulate were proposed, but in every case it turned out the person had accidentally used the parallel postulate at some critical step in their argument.

The question remained open for two millennia until the 1800s, when an obscure Russian discovered what the problem had been all along. The parallel postulate is not provable, because there are perfectly self-consistent "non-euclidean" geometries where it breaks down. All of the other axioms hold fast, but the parallel postulate no longer applies. The difference is that normal Euclidean space is "flat", but non-euclidean spaces bend and twist. In fact, the parallel postulate is the defining feature that sets flat space apart from other geometries.

The best example of a non-Euclidean geometry is the curved surface of the earth. If you try to walk in a straight line on the earth's surface, you'll end up looping all the way around it in a giant circle like the equator. These great circles are the "straight lines" on the surface of a sphere, and any two of them will intersect. So there is no such thing as "parallel lines" on a globe!
If you and a friend line up side-by-side along the equator and start walking due South, i.e. parallel to each other, you'll bump into each other in Antarctica. The straight lines are only "parallel" for a brief moment as you start walking, and after that the curvature of the space makes them converge toward each other and eventually collide! The earth is said to have "positive" curvature; in a space with negative curvature you and your friend would diverge and get farther apart. Only in a flat space, with curvature 0, will you keep the same distance.

In the early 20th century Albert Einstein discovered that even normal space, the good old three dimensions that we live in and that Euclid thought he was studying, is itself twisted and warped like the surface of a globe. The curvature is very slight, at least within our solar system, so we don't notice it in daily life. On the scale of galaxies though, or near massive objects like a black hole, Euclidean geometry goes out the window.

The curious history of the parallel postulate raises an interesting question: are the axioms actually true? I mean, we accept them without question for the sake of a particular proof, but are they actually true in any objective sense? Broadly speaking there are three answers people give to this question: "yes absolutely", "yes empirically" and "sure, hypothetically".

The "yes absolutely" crowd is, of course, the Platonists. For them numbers and shapes are real entities, dwelling in the Realm of Pure Forms outside of time and space. Axioms and theorems are just properties that these Forms have. If the fabric of space and time is warped, that just means that the physical world isn't as tightly coupled with the Form of Flatness as we thought. For Platonists, axioms aren't any more fundamental than theorems - that would be like saying an apple's sweetness is more fundamental than its texture. Axioms and theorems are all just true facts about a real object, and it's an academic exercise to figure out which of these facts can be logically deduced from others.

The people who say that axioms are "yes empirically" true treat math as a branch of science. Axioms aren't absolute, heaven-sent truths. They are just empirical facts about the world that have been so reliable that we're willing to put some faith in them. The parallel postulate is a great example of this: we relied on it for millennia, and used it as a first principle in science and engineering. With the help of the parallel postulate, our buildings have stayed upright and our crops have been harvested. Einstein showed that, strictly speaking, the parallel postulate is only approximately true for the world we live in. However, it is such an excellent approximation that it still works well enough for everyday use. By and large the ancient Greeks (Plato and Pythagoras being conspicuous exceptions) seem to have taken this attitude toward mathematical axioms. For them, geometry was a science on par with physics and biology. If Euclid had been able to get into a space ship and travel to a black hole, he would have discovered that the parallel postulate didn't always hold, and maybe he would have revised The Elements.

Finally, there are those who say the axioms are "sure, hypothetically" true. This school of thought treats proofs just as exercises in logical deduction. When you prove a theorem you aren't proving it about any specific entity, not even a purely abstract Form. For that reason it's not even meaningful to ask whether your axioms are true. Instead you are doing a purely hypothetical proof, applicable to any situation where your axioms *happen* to be true. Again I'll use Euclid as an example. He thought he was studying the geometry of flat space, and the idea of non-flat geometries never occurred to him. However, whenever he wasn't using the parallel postulate he was inadvertently proving theorems about curved space as well. His construction of an equilateral triangle that we saw, for example, works perfectly well on the surface of a globe. You could use his construction to chart out a triangle connecting Seattle, New York and some place in northern Canada. If Euclid had wanted to write a version of Elements for curved spaces, he could have just copy-and-pasted large swaths of text from the original, since the proofs carried over exactly. This "sure, hypothetically" school of thought doesn't concern itself with what is true or false. Instead, it focuses on the mechanics of deductive reasoning.

Personally I'm not married to any of these interpretations. They all have their merits. Even Platonism is defensible, if you recognize that the Realm of Pure Forms is located between your ears, and the Forms themselves are artifacts of human cognition. It is true, after all, that a "perfectly round circle" is a hypothetical idea that all people seem to agree on. Whichever school you subscribe to though (if any), math is fundamentally a linguistic exercise.
Whether you're talking about the Form of Triangles, a crude triangle drawn in the dirt, or just some hypothetical three-sided shape, you're still *talking*. And anything that you can talk about you can also reason deductively about. You can answer philosophical questions about the nature of language however you want, but your answers will carry over to math too. Perhaps, in seeing this connection, Plato was right after all.

5 MATH AND SCIENCE

It's always seemed weird to me that in so many people's minds math and science go together like peanut butter and jelly. Like Harold and Kumar. Like homework and procrastination. They don't. They never have. They probably never will. Math was invented mostly to do commerce, agriculture, civil engineering and astronomy, like we saw in chapter three. Astronomy kinda counts as science, but at the time it was mostly used as a tool for religion and agriculture. Even up through the present day, most math is done in the context of economics, logistics, and more recently computers.

When people think of a scientist the image that often comes to mind is a white-haired Einstein, scrawling obscure formulas across a chalk board. But people like that are only a tiny fraction of the scientific community, and they are only able to exist because other scientists spend most of their time down in a lab or out in the field.
A better picture of a scientist is Charles Darwin trekking across the Galapagos Islands, carefully recording every variety of finch he sees and trying desperately to keep his field notes dry. Or it's Marie Curie holed away in her lab for days at a time, painstakingly distilling radioactive isotopes from raw ore. It's grimy, tedious, thankless work - and depending on your temperament, it's the most exciting thing in the world. Some scientific theories, like gravity, are best expressed using equations. Others, like evolution, are not.
Science isn't inherently about head-in-the-clouds abstractions.
It's about getting down in the mud to try and figure what makes some little part of the universe tick. And in this country it's also about applying for government grants, but that's another story...

With that disclaimer out of the way though, the story of math and science is well worth telling. Partly because it's so much better documented than the Great Leap Forward or the rise of civilization. We actually know who invented what - in some cases we even have their original manuscripts on hand. We know how it challenged prevailing notions about the way the universe operates, and we know what it offered as an alternative. The scientific revolution changed the world forever, and math was a key part of it.

About 450 years ago, a young English schoolboy named Isaac was in a tough spot, the kind of tough spot that has plagued many a schoolchild over the ages: bullies. There was one bully in particular who was terrorizing the class, and seems to have especially had it out for Isaac. Besides trouncing the other students in physical fights, this particular bully added insult to injury by being the top student in the class (which perhaps got him a certain leniency from the school authorities). Poor Isaac, in contrast, was shrimpy in size, mediocre in his studies, had very few friends, and was socially awkward. It was not a good situation.

One day it reached a breaking point, when the bully sucker-punched Isaac in front of the whole class during recess. Isaac fought back in desperation, to defend himself and possibly to salvage a little bit of honor. To everybody's surprise, it quickly became apparent that Isaac was the better fighter. As the rest of the class cheered, Isaac thoroughly defeated the bully, and topped it off by rubbing the boy's face in the mud.

That wasn't enough though. This bully seems to have gotten particularly far under Isaac's skin (or, perhaps more likely in light of later events, Isaac was one to hold a grudge). Being beaten up in front of the whole class wasn't nearly degrading enough, so Isaac took it upon himself to humiliate the bully in every area, most notably academics. He threw himself into his studies with the same ferocity that had served him so well on the playground, and quickly displaced his enemy at the top of his class.

That bully may have done the world a huge favor. Isaac didn't have a very promising future. Socially awkward, academically mediocre, and the son of a poor widow, he seemed destined for a low-key career as a farmer. However, the bully showed Isaac that, if he applied himself, there was no limit to what he could do.

As you may have guessed by now, Isaac's last name was Newton. He became the single greatest scientist, and arguably also the greatest mathematician, in the history of the world. More relevant to our purposes here, he was also the one who made physics into the first truly mathematical science.

Newton's work in the 1600s marked the high point of the Scientific Revolution, but the revolution had its roots in the philosophers of ancient Greece. Books like the one you're reading generally focus specific individuals who changed the world, partly because it makes for a better story. In reality most great science comes from a vibrant community, with lots of ideas being swapped back and forth. It's easy to forget that Alfred Wallace discovered evolution at the same time as Charles Darwin, or that Henri Poincare had figured out 90% of special relativity by the time Einstein put his stamp on it. Newton himself famously stated that "if I have seen further, it is by standing on the shoulders of giants". In the same way the idea of empirical science was the product of many minds, but one of them stands out above all the rest: Democritus.

Democritus lived between the times of Thales and Euclid, in the North of Greece. As a young man he received a large inheritance, and spent it traveling the known world. He travelled throughout Greece. He went to Egypt and lived there for five years, learning everything he could about their culture and philosophy. He may have even looked up some of Ram's old works, which were over a millennium old by that point. It looks like Democritus even got as far as Ethiopia and India, meeting with every great thinker he could find and absorbing all knowledge he could get his hands on.
When his wanderlust was finally sated, he came back to Greece and devoted himself to natural philosophy.

The main reason Democritus is famous today is that he was the first person to suggest the world was made of atoms (though his conception of them was quite different from modern notions). However, he was also the person to really lay out the scientific method. For him knowledge came in two types, which are often translated from Greek as "bastard" and "legitimate". Bastard knowledge is what we see and touch in the world, and how it appears to us. Legitimate knowledge, on the other hand, is only perceived by the mind. It's the underlying patterns that govern the world, and it must be found inductively by observing the world and reasoning about it.

Plato made a similar distinction, but for Plato the legitimate knowledge had no connection to the real world. It was about the Realm of Forms. Rather than reasoning from observations to see general patterns, as Democritus advocated, Plato wanted to cut the real world out of the loop.

Plato, perhaps not surprisingly, loathed Democritus. He hated him so much, in fact, that he allegedly advocated burning all of Democritus' books. In the constant ebb and flow of history, I'm sorry to say that Plato eventually won out. Everybody today learns about Plato in school. He's even made it into our language, with words like "platonic". You'll generally only hear about Democritus if you study philosophy or the history of science.

As Greece eventually declined, Rome rose to power, and finally the middle ages began, the spark of empiricism gradually faded from Europe. By the middle ages, "science" was based on the ideas of Plato's student Aristotle. Aristotle's ideas had been adopted by the Catholic church not as scientific hypotheses, but as party-line dogma. Aristotle and a few other Greeks had figured out the big picture of how the world works, except of course for additional knowledge revealed by the Christian religion, and their teachings were the starting point for any further inquiry. End of story.

According to Aristotle the world was divided into the "earthly" realm and the "heavenly" realm, each of which was governed by different laws. In the earthly realm, the natural state of matter was to fall to the ground and come to rest. It was a gritty, imperfect place, having long ago been polluted by the Original Sin of Adam and Eve (the part about original sin comes courtesy of Catholic theologians, not the pagan Aristotle). The heavens, in contrast, were filled with perfect luminous spheres. Here matter moved forever in elegant circles, untainted by the crude world below.

Now in fairness I should note that Aristotle himself was a legitimate scientist. He prized empirical observations, and disagreed with a lot of Plato's metaphysics, including the Realm of Forms (which even Plato may have abandoned later in his life). His theories about physics ended up being wrong, but they were very reasonable ideas at the time, and his work in biology was first-rate. I'm sure Aristotle would have been fascinated to learn where his science had missed the mark, and would have changed his views when the evidence mounted up. However, the powers that be in medieval Europe were not so open-minded, and the Scientific Revolution was largely about finally showing that Aristotle's theories, which had stood for over a millennium and been embraced by the church, were wrong.

The first great blow came from Copernicus, who suggested in the mid 1500's that maybe, just maybe, the Earth orbited the Sun and not the other way around. As astronomers began tracking the stars and planets in greater detail it became clearer and clearer that Aristotle's geocentric model just wasn't up to snuff. The idea of the planets revolving in perfect circles simply didn't match up to observations, and astronomers were force to postulate additional movement on smaller circles to explain the gap. Even that didn't really cut it. In contrast, everything worked out beautifully if all the planets moved in ellipses around the sun.

The astronomical data was being gathered by men like Tycho Brahe and Johannes Kepler, but the real bane of Aristotelian physics was Galileo Galilei. Galileo, who coincidentally died the year of Newton's birth, was the first man to use a telescope to study the stars, taking a closer look at the supposedly pristine heavens. He found that the sun had irregular blotches, now known as sun spots, marring its face. The planet Jupiter had streaks and even an oblong red splotch. Most shocking of all, the planet Jupiter was itself orbited by 4 additional worlds.

What is Math?

Galileo also challenged Aristotle in the terrestrial realm. He suggested that the "natural state" of moving objects was to just keep on moving under the power of their own inertia, rather than coming to rest as Aristotle had decreed. They only come to rest because of air resistance and other external forces.

By the time Newton entered the stage it was pretty clear that Aristotle's cosmology was fundamentally flawed. What was not clear, however, is what might replace it.

The story goes that one day Newton was sitting under an apple tree, thinking about the way that moons orbit around their planets and planets, in turn, orbit the sun. All of the motions were in clean, elegant ellipses, a strong hint that the same mechanism was behind all of it. As he sat lost in thought, an apple fell from the branches above and clocked him on the head, and he had a brilliant realization. The force of gravity that pulled the apple toward the ground could conceivably reach out to the moon as well. What would it look like if that was the case? Well, the moon was already moving rapidly through space as it orbited the Earth, so the pull of gravity wouldn't just pull it to the ground like a falling apple. Instead, gravity would pull it off its course, making it veer away from its current path in an arc toward the earth. In orbiting around the Earth, the Moon was already moving in just such as arc.

If you throw a baseball it will travel a few yards, then gradually arc down toward the Earth and hit the ground. Throw it harder and it will go further. If you ignore air resistance for a minute, with enough force you could conceivably throw the ball many miles, so far that it travels half-way around the earth before finally landing. Of course nobody has that kind of arm on them, but it could be done in principle. Now imagine throwing the ball so hard that it loops all the way around the Earth and comes back at the level of your head (this would require the ball to be moving at 25,000 miles an hour - quite a throw!). At this point the ball is no closer to the ground than when it first left your hand, and it will still be moving at blazing speed. In theory it will just keep looping around the

Earth forever, always "falling" but never hitting the ground. This is exactly like the orbits we see out in space! The Moon was essentially a giant baseball, thrown by cosmic forces into orbit around our planet. Newton realized that the force of gravity can explain all of it, uniting the Earthly and Heavenly realms into one system of physics.

It's not clear whether the famed apple ever actually fell on Newton's head, but if so we know when it happened. In 1665, the university temporarily closed due to fears about the bubonic plague. Newton headed back at the family farm to wait out the break. He had to find something to do with his time, and his hatred of farm work was matched only by his incompetence at it, so his thoughts drifted to physics. During this impromptu break from school Newton worked out a revolutionary new theory of physics, breaking down the heaven/earth divide that had dominated science for over a millennium. The mathematics of the day wasn't able to describe his physics, so he also invented a whole new branch that is now called calculus. Then what did he do with all of this revolutionary material? He put it in a drawer for the next several years and told nobody. I kid you not.

It was only later, when other scientists began edging closer to some of the same ideas, that Newton finally started publishing his work. Newton's, um, "colorful" personality streak, the one that drove him to humiliate the schoolyard bully back in his childhood, matured into a vicious paranoia about other people getting credit for his discoveries, even if he had never bothered to tell anybody about them. Newton's brilliance as a scientist should not be confused with functionality as a human being.

Galileo is the scientist who broke the back of Aristotelian physics, but it was Newton who created a replacement for it. Newton's work remained the best available description of the universe until 1905, when Einstein published his paper on special relativity. Even today, for practical purposes it's almost always better to use Newton.

In an echo of Euclid's axiomatic approach to proofs, Newton outlined three laws of motion, expressed in a combination of English (well, Latin actually) and math. These three laws, plus a formula for the force of gravity, were Newton's first principles, analogous to the axioms of geometry. Using those laws you can derive everything from how billiard balls move on a table to the shapes of planetary orbits. Newton ultimately laid out his new math and physics in Principia Mathematica Philosophae Naturalis (Latin for Mathematical Principles of Natural Philosophy), a book which is to physics what Euclid's Elements is to geometry.

It's difficult to over-state the importance of Newton's work. On the one hand, he put the final nail in the coffin of Aristotle's physics, closing the door forever on almost two thousand years of dogma. He also united earth and the heavens into one realm of physical laws. Maybe most importantly, Newton gave the first coherent, mathematical account of the world starting from first principles. This idea of a "clockwork universe" became a bedrock idea for the Enlightenment.

Newton's laws are actually one of the best examples where math and natural language blend together. They can be expressed either in equations or in normal, everyday language. Here are Newton's laws as you've probably seen them before, in English:
1) An object at rest tends to stay at rest unless something pushes on it. Similarly, a moving object keeps moving, with the same speed and direction, unless an external force pushes it.
2) When an object gets pushed it accelerates in the direction of the push. For a given push, heavier objects accelerate less than light objects.
3) For every action there is an equal but opposite reaction
The first law is the principle of inertia, the one that flies in the face of the Aristotle's heaven/earth divide. Aristotle thought that motion was the natural state of objects in the heavens, but resting on the ground was the natural state of objects on Earth. The law of inertia says that in both realms the natural state is constant, straight-line motion (or lack thereof). The reason we generally don't see this "natural motion" is that in both realms there are other forces at play. In the heavens the moons and planets are pulled by

the force of gravity, and on the earth there's also the force of air resistance. The second law just says that if something *is* pushed by an external force (like gravity or air resistance), then it accelerates in the direction it's being pushed. An object that is twice as heavy will accelerate half as much. Finally, the third law says that if thing A pushes on thing B, then thing B pushes back equally hard. This is why, if you lean on a wall, the wall pushes back on you and keeps you upright.

If you want to break out the mathematical notation, these three laws boil down to two equations:
$(BFBA)/mA = aA$
$FAB = -1*FBA$

The first equation captures laws 1 and 2. For an object called A, if you add up all the forces that other objects are exerting on it and divide by A's mass, you will get its acceleration. If it turns out that there are no forces, then the acceleration will be zero and the object will continue moving (or staying still) just as it is, i.e. the law of inertia. The second equation says that two objects will push each other equally hard but in opposite directions. That "push" could be attraction from gravity, blunt force from running into each other, electromagnetic repulsion, or anything else. This law just says that, no matter what the force, every object gives as good as it gets.

Physics was the first of the sciences to be reduced completely to math, and because of this they became closely intertwined. Calculus was developed mostly as a tool for manipulating the equations above and applying them to new situations, and ever since then math and physics have had a symbiotic relationship. Physics poses problems that can't be solved with existing math, so new methods are invented. Mathematical discoveries are made, which cast light on existing theories of physics.

Several years ago I visited Westminster Abbey in London, where Newton is buried. For a few brief moments in time I was, out of all the people in the world, the closest one to Newton's earthly remains, and I was surprised to find that it was almost a religious experience. I've devoted much of my life, personally as well as professionally, to mathematical science, and I was in the presence of its founder. Out of all the stars in the universe of science, Newton's shines the brightest.

There were a lot of surprising things about Newton, both good and bad. He was a mediocre student, but ended up being a world-class genius. He was paranoid and held deep grudges. He probably died a virgin. But the most surprising thing may be that, in his own mind, Newton's most important work wasn't science at all. It was religious scholarship.

Newton was a devout Christian, and he studied religion with the same dogged obsession that yielded such fruits in physics. He actually wrote more on religion than he ever did on the natural sciences. However, the details of his beliefs were extremely unorthodox in a time where heresy could be a capital offense in England. For that reason scholars are still debating the details of his views, since he couldn't exactly say them publicly. Most of his work focused on Bible translations (he found a number of small errors) and historical dating. He estimated that Jesus was born in 33 C.E., which jives with the best estimates available today, and he even tried his hand at predicting the end of the world. According to his calculations the world will last at least until the year 2060, and he concluded by saying
"This I mention not to assert when the time of the end shall be, but to put a stop to the rash conjectures of fanciful men who are frequently predicting the time of the end, and by doing so bring the sacred prophesies into discredit as often as their predictions fail."

Does that sound familiar?

Newton's work may have been a "first principles" approach to physics in the same way that Euclid's Elements is to geometry, but the similarity stops there. Remember, Euclid was taking a bunch of things that people already knew and trying to put them on a more solid foundation, but Newton was breaking genuinely new ground. As such he had no patience for logical nit-picking. There were discoveries to be made, and not enough time to cross every t and dot every i! In fact, the first time Newton ever looked at Elements he concluded that it was just proving things that were already obvious, and disregarded the whole book. He wanted knowledge, not legalistic rigor! Newton's math ended up being a hodgepodge of rigorous algebra, suggestive pictures, and assertions that he thought were "obviously true". It took mathematicians centuries to hammer out all of the details in a way that would have satisfied Euclid, and in the end they only confirmed that Newton had been right all along.

If you hang out with physicists, or engineers, or computer scientists doing math - really anybody outside of a university math department - you'll quickly notice that they aren't operating in a rigorous way either, any more than a professional author studies the theory of English syntax. It's a much more organic process, typically starting with a hunch or a picture in somebody's mind. It then progresses through a lot of head scratching, some algebra, and a peppering of intuitive leaps that would make Euclid shudder. Physicists are so notoriously cavalier that we even have a name for these conspicuous gaps in the rigor: hand-waving.

Imagine a physics professor at the board in the middle of a long algebra derivation. At some point she stops, and begins to stumble over her words a bit. "For the next step, you can picture what this would look like physically" she says, and sketches out a crude drawing. "If you imagine the length of the string being small enough, you should be able to see that...". As she's talking she starts gesticulating toward the picture, or even just wildly into the air (hence "hand-waving"). The blunt fact is that she doesn't have a rigorous proof of what she's saying, but to her it is obviously true, and she's trying to convey that intuition. This is not an axiomatic proof, but it's how physicists roll.

Hand-waving is not the same thing as arguing by bold assertion. Quite to the contrary, the key thing that makes hand-waving possible is having a deep familiarity with the mathematical tools at hand, one that runs deeper than just being able to recite theorems or chug through algebra. In the same way that you don't start from theoretical first principles when dealing with a problem in your daily life, hand-waving in math is based on having well-honed intuitions about the subject. You don't just know it; you understand why it works the way it does. Equations you scrawl on the board are based on pictures in your head, not the other way around.

This familiarity amounts to a broadening of our intuitive scope. It can be done! Nature has furnished humans with the number sense and other basic instincts, but it's also possible to train new ones. Or at least, we can adapt existing instincts to master new ideas. Broadening your intuitive scope can take a lot of work, but there's nothing quite as satisfying as banging your head against a thorny concept for hours and days, only to have the scales suddenly fall from your eyes and see that it all makes perfect sense. That cycle of agony and ecstasy is what has always kept me coming back to math.

Euclid guarded against incorrect results by requiring that every step in a proof be scrutinized for validity and traced back to first principles. Physicists guard against incorrect results by demanding such a good intuition for the math that they can usually tell when they're wrong. On the rare occasions when an error does slip through the cracks it can usually be caught by another physicist or, in the worst (and very, very expensive...) case, by an experiment that goes awry.

Often physicists will invent a new technique, use it for years, and already be on to the next subject by the time pure mathematicians have fleshed out all the details. These extra details shed a lot of light on when the tool breaks down, and occasionally they even suggest new or more general ways to look at it. But they almost never find an actual problem in the physicists' conclusions.

It was fascinating for me to learn all this as an undergraduate. I majored in physics, and took a lot of computer science and mathematics on the side. Math classes beyond calculus are based almost 100% on precise definitions and rigorous proofs from axioms. A typical homework assignment won't be a bunch of problems to solve so much as theorems that you need to prove. In contrast to this, they don't even teach the axiomatic method to physics undergrads. Computer science students see a little bit of it, but usually just as a highly specialized sub-discipline. Physics and computer science do copious amounts of math, but they understand that well-trained intuitions and convincing "picture proofs" are more natural than rigorous axiomatic derivations, and they are the key to new discoveries and applications. Humans didn't evolve to use any language in isolation: there is supposed to be a constant interplay between the language we use and the concepts it describes.

Newton's laws of motion are not self-evident, especially considering how many great thinkers have disagreed with them. However, they're not implausible either. A billiard ball will stop quickly if you roll it in mud, but will travel a long way on a smooth table. Is it really such a stretch to imagine that it would keep going forever on a perfectly smooth surface? Similarly, it is obvious that pushing on something will make it move, with bigger objects moving less for the same push. And you only need to push on a wall to see that it pushes back on you. If you squint your eyes and look at it just right, it's almost as if Newton's laws are an everyday understanding of the world that has been reduced to equations. Of course it only looks this way with the benefit of hindsight, yet the fact remains: even if solving the equations is difficult, the physics being described is intuitive. But alas, that is not always the case.

In doing reductionist science we're generally aiming to make sense of the world. We peel back layers of the onion of knowledge, trying to bring more of the world within our realm of understanding. In the 20th century though this process back-fired. We discovered that Newton's theories are only approximately true. They work great for billiard balls and planets, but when you try applying them to small things like atoms, or large things like galaxies, their predictive power breaks down. What displaced Newton threw a monkey wrench into the whole notion of humans actually making sense of the world. It's a stark reminder that the interplay between human intuitions and mathematical equations must always be a two-way street.

It turns out that the largest structures in the universe are best described by Einstein's theory of relativity, which we referenced in the previous chapter. Einstein discovered that Newton's understanding of gravity was incomplete. Gravity is not just an attractive force between objects, but a warping of space and time itself. What looks to us like a force is really just objects moving in straight lines through curved space and time. In Einstein's world space and time blur together into a single, bizarre entity, which bends and even ripples in response to matter and energy. My friends tell me that with years of practice it is possible to understand relativity intuitively (I can't attest to this personally; I don't understand it), but the only way to do it is by molding your brain around the equations.

Quantum mechanics, the rules that govern the tiniest objects in the universe, is even worse. Most people agree that while it's possible to get to a point where you understand relativity, quantum mechanics is simply incomprehensible. The human brain just isn't up to the task. Once you have the equations in hand you can develop an intuition for how they behave. Considered as strictly mathematical objects, the equations of quantum mechanics are actually fairly simple, and often very easy to solve. But the equations don't describe anything that can be envisioned physically. In fact, the physical picture they paint is absurd, and physicists would laugh quantum mechanics out of the room except for the fact that it's the most tested theory in the history of science and there's no good alternative. Some of the greatest minds of the 20th century have labored in vain to paint an intuitive picture of the quantum world, and it has come to nothing. Quantum mechanics might be fundamentally and forever outside of our intuitive scope.

From a linguistic perspective quantum mechanics presents an interesting conundrum. We have a truly nonsensical description of the world that just happens to have extraordinary predictive power.

Is it even a "description" in that case, if we can't understand it? This isn't Euclid's problem, where human intuitions were polluting the pure mathematical logic. There *are* no human intuitions here - we're flying blind. The only thing tethering quantum mechanics to reality is the predictions it makes about the outcome of experiments, and so far so good.

What should we make of all this? It might just mean that there's a more sane theory underlying quantum mechanics, and we haven't been clever enough to find it yet. Einstein spent the last part of his life searching for such a theory, with no success. It could also be that the fundamental principles of the universe simply aren't conceivable to the human brain. When you think about it, even getting as far as we have (subatomic particles, the age of the universe, etc.) is pretty impressive for a race of bipedal apes.

Aside from possible fundamental limitations to human comprehension, there is a much more mundane problem with relying on your intuitions: it's very tricky to tell just how solid they are. I got an embarrassing lesson in this fact a few years ago when I was studying the performance of computer hard disks.

My collaborators and I were trying to derive a formula for how long, on average, it takes a computer to read memory off a hard disk. We were using a drastically idealized model of how these disks work, similar to models that other researchers had used in the past. Despite all the simplifications though, the problem turns out to be extremely subtle. A number of first-rate mathematicians had tackled the problem, gotten their work past peer review, and published results in prestigious journals, only to find out later that there was an error in their derivations and their formulas were all wrong.

Now I was a young male at the time and hence, by definition, smarter than all those who had gone before me. I tried my own hand at the problem, plugging a few of the more glaring holes in previous work. I embarked on a long derivation, with several points of very conspicuous hand-waving, and eventually arrived at a formula of my own. My formula was very complicated, so you couldn't exactly look at it and say "oh! I see now why it works this way!". It was more of a grotesque black-box. However, it seemed reasonable as near as I could tell, and I was confident in my derivation, so I started writing a paper to present my findings to the world.

My co-authors were a graduate student and a professor. Both were brilliant mathematicians, and they had both looked at my derivation and said it seemed correct. The professor though was very cognizant of this problem's history, and she insisted that we simulate our model using a computer to see, empirically, what my formula should output. You know, to cover the one-in-a-million chance that I had made a mistake...

As you've probably guessed by now the simulation didn't turn out as hoped. It was very close, but there was a small gap between my prediction and the computer's output, and it only got more and more definite as we ran additional simulations. Eventually I faced facts and went back to the drawing board to figure out what had happened. Lo and behold, it turns out I had botched the math in essentially the same way as the people I thought I'd out-smarted! I had made fundamentally the same mistake, but just swept it further under the rug than they had. No part of my derivation could be salvaged. Back to square one.

Now officially out of ideas, I went back to the (damn) computer simulation, used it to produce a bunch of data points, and sat down trying to fit something to the data. I wasn't doing a derivation or a proof mind you - just throwing formulas at the wall and seeing if anything stuck. Eventually I stumbled upon one that was stunningly simple and fit the data like a glove. No way that could be coincidence! Filled with renewed zeal I sent out a flurry of excited emails, and then proceeded to stare blankly at the magical formula for several hours.

Eventually a new idea dawned on me, and I took a totally different approach to the derivation. This one worked. We were finally able to publish the paper, and nobody would ever be the wiser. Wait, on second thought why am I telling you this?

I'm not the only person to have received a humbling lesson in mathematical rigor; it happens to everybody now and then. The interplay between intuition and rigor is a constant balancing act.

Even the great Newton himself, who scorned the Elements in his youth and relied so heavily on hand-waving, developed a deep appreciation for Euclid later in his life. At one point he went so far as to write an appendix for an edition of Elements, which I quote here:

> Independent, however, of the many other advantages which a knowledge of [geometry] affords to society, at large, the study of it greatly tends to strengthen and improve the mind; inuring it to the habit of close thinking and just reasoning.
>
> Hence many valuable treatises, on this subject, have been written, and considerable advancements and improvements have been made, in this useful and pleasing department of science. But of all the systems of geometry, ancient or modern, that have hitherto come to our notice, none has a higher claim to distinction than that which is known by the title of Euclid's Elements. This may indeed be considered as the basis for all our best modern productions, and the true standard of taste and elegance in geometry.

Every mathematical person has a miniature devil sitting on one shoulder and a miniature angel on the other, both whispering in their ears. Their names are Euclid and Ram. Which one is the devil and which is the angel depends on the context and personal temperament. Ram says "close your eyes and picture it. That makes sense right? Draw a diagram, think hard about what the solution has to be, and then write down the answer. You can do it!". Meanwhile, Euclid constantly frets "umm, you should probably double check that", wringing his wrists anxiously. Ram is the idea guy, the one who keeps his eye on the big picture and leads us to all the cool new discoveries, but Ram can also be reckless and foolhardy. Sometimes, even if he can be a stick in the mud, it pays to listen to Euclid.

6 THE TRUE, THE FALSE, AND THE PATENTLY ABSURD

I'm certainly not the first person to wonder about the nature of math. A lot of people have said quite a bit on the subject, going as far back as Plato suggesting that numbers lived in the Realm of Pure Forms. For most of history, philosophizing about math has always taken a distant back seat to actually *doing* math, but that changed for a period in the late 19th and early 20th century. Some logical holes were discovered in the foundations of mathematics, sparking a heated debate on the nature of math and a quest to put it on solid foundations. People didn't want to answer these questions just with additional philosophizing. Like true mathematicians, they wanted exact answers with rock-solid arguments to back them so. So like a snake eating its tail, mathematics was turned upon itself as an object of study, with startling and controversial results. In this chapter I will give an overview of this story, how it started and ended, and the ripple effects it has had up through the present day.

You can skip this chapter if you want. It's not necessary for understanding the rest of the book, and it gets into some pretty abstract (and a bit trippy) topics. On the other hand, if philosophy still interests you even when it's not very applied, you may well find this to be the coolest part of the book. The battle over the foundations of math is a fascinating study in just how far we can push the limits of language.

Spoiler alert! One of the coolest duels in the history of Western movies comes at the end of The Good, the Bad and the Ugly. It's a three-way showdown between the titular characters. The "good" one is a nameless bounty hunter played by Clint Eastwood. He is stoic and cocksure, and not above being morally dubious when it's expedient. I'd say he's an anti-hero at best, but I guess he's a good guy by spaghetti Western standards. The "bad" is Angel Eyes, a ruthless mercenary with a real zest for his murderous profession. Finally, Tuco Ramirez is the "ugly", an uncouth, general-purpose scoundrel. He doubles as the bumbling comic relief of the film, but put a gun in his hand and he's every bit as deadly as the other two. The movie's climax has them in a tense three-way duel over a stash of buried gold. Each gunslinger would like the other two dead, and the question is who will shoot first, and at whom. I won't spoil the ending, but suffice to say it's a great movie!

In the early twentieth century, mathematicians the world over were in a similar situation. The confrontation wasn't about hidden gold, but about the meaning of mathematics. Nobody got their brains blown out with a shot from the hip, but careers and lives were ruined, and academics fought a venomous war of words that spanned decades. In the first corner were the Intuitionists. They were a motley group who varied a lot in their specific attitudes, but generally they agreed that math is a human invention. Some of them went so far as to question the laws of logic itself. For the intuitionist, numbers "exist" only insofar as somebody is thinking about them.

The second group was the Platonists, who I've discussed previously. They were the intellectual heirs of our favorite cult leader Pythagoras, and claimed that math is the pursuit of transcendent truth. In their minds numbers and shapes were real entities, and mathematics studies their properties in the same way that physicists study stars. As with their Greek forebearers, the Platonists' beliefs often blended into religion and mysticism.

In the last corner you had the Formalists, who were by far the newest school. They threw up their hands when it came to the meaning of math. For them it was all just a game of symbols, much like the context-free grammars I discussed in chapter two. There were no deeper truths to be found or philosophical stones to unturn. The formalists didn't worry about whether numbers and shapes were real or abstract. They weren't even particularly concerned with true versus false. For them it was all about what statements you could prove by following the rules of the symbol game.

Each of these schools had its own strengths and weaknesses, and each had its luminaries. Intuitionism boasted Henri Poincare, a universal genius who arguably beat Einstein to the theory of Relativity. Formalism was founded by David Hilbert, who set the tone for 20th century mathematics and laid the groundwork for quantum mechanics (although it's not 100% clear that Hilbert himself would count as a formalist). The Platonists included Georg Cantor, who started the whole hullabaloo, and Kurt Godel, who brought it to a dramatic close.

If you're like most people you've never thought too hard about the distinctions between intuitionism, platonism and formalism. They probably never even occurred to you. I mean, why would they? It's not like these schools of thought would get different answers to algebra problems, or disagree about how to balance a checkbook. You could make a strong argument that all of this bickering was academic navel-gazing at its worst, and if that's how you feel then you're not alone. Aside from the woo-woo elements like Pythagoras, most mathematicians up through the 1700s would have agreed with you that the whole argument was a colossal waste of time. Math was something they worked hard to build up, but they never lost any sleep over digging deeper.

However, that all changed in the late 1800s and early 1900s. Esoteric questions, like the nature of infinity and what constitutes a proof, suddenly became hot button issues, and mathematicians were forced to take philosophical sides. The trouble started with a new topic in math called "set theory", a rigorous way to study infinity, and the aforementioned German mathematician Georg Cantor. However, the roots of the issue go back to the scientific revolution.

Galileo Galilei was a scientist in Italy during the 1500s and 1600s, who I mentioned in the last chapter. A lot of people consider him the father of modern science, since he was the one who, more than anybody else, pioneered the physics experiments and astronomical observations that kicked off the Scientific Revolution.

What is Math?

Physics in Galileo's time was built up on the ideas of the Greek philosopher Aristotle, whose views had been adopted by the Catholic Church. According to Aristotle, the world is divided into the Earthly and the heavenly realms. On Earth the nature of things is to come to rest. If you throw a ball it will land, roll, and eventually stop. If you throw it harder it will fly longer, but the law of inevitable rest will catch up with it. In the heavens, by contrast, things are a lot cleaner. The sun, stars and planets are all perfect spheres, floating forever through the cosmos in great, elegant circles.

Galileo had the bright idea to point a telescope at the sky, and the world changed forever. He discovered that the planets are not pristine spheres. Mars has valleys and dry seas. Jupiter has a big red spot, and is orbited by moons. Galileo found himself in hot water with the church and spent the rest of his life under house arrest for heresy. That sucks, but at least it's better than what happened to his contemporary Giordano Bruno, who had the audacity to suggest that the Sun was just another star. Bruno was burned at the stake.

Galileo wasn't really a mathematician. He was proficient in math by the standards of his time, but for him is was just a tool he used for doing science. Unlike Newton, who ended up inventing calculus to solve physics problems, Galileo's science didn't require any new math, and he just made use of the tools available to him. However, Galileo did notice a little mathematical curiosity, dealing with the perennially-mysterious subject of infinity.

There are infinitely many natural numbers. One, two, three, and so on. You can multiply any of these numbers by itself to get its perfect square, as shown in this table:

Number	Perfect Square
1	1
2	4

3	9
4	16

Some natural numbers, like four, are also perfect squares, but most aren't. As you can see in the table only a fourth of the numbers up to sixteen are perfect squares. Up to one hundred it's only one tenth. Only a measly 1% of the numbers below 10,000 are squares, and the ratio keeps dropping there. There are a LOT more natural numbers than perfect squares!

Now imagine that I expanded the table above to cover all natural numbers. There would infinitely many rows, one for every number, and each number would be paired up with its square. There would not be any duplicate entries in either column, so the table would give us a one-to-one matching between the set of all natural numbers and the set of all perfect squares. So far so good?

Here's the rub. If you can put two collections in one-to-one correspondence, then that means they have the same number of elements. For example, if you can match up the fingers on your hand with the cars in a parking lot then that means there are five cars, just as you have five fingers. So it seems that there are as many natural numbers as there are perfect squares.

Ok, we have a paradox here. There are obviously fewer perfect squares than natural numbers, since only a diminishing fraction of all numbers are perfect squares. But on the other hand, if you can line up two collections perfectly one-to-one, it seems like they have to be the same size. So are there more natural numbers than perfect squares, or are there equally many? Or is that even the right question to ask?

Galileo noticed this conundrum, and concluded that the very question is off limits. There are infinitely many natural numbers and infinitely many perfect squares, and that's all there is to say on the matter. Neither collection is "bigger" than the other, because the notions of "bigger", "smaller" and "same size" don't apply to infinity. The reason this is so hard to accept is that we humans are finite creatures, and our intuitions are all based on a finite world. To quote Galileo:

> We can only infer that the totality of all numbers is infinite, that the number of squares is infinite, and that the number of their roots is infinite; neither is the number of squares less than the totality of all numbers, nor the latter greater than the former; and finally the attributes "equal," "greater," and "less," are not applicable to infinite, but only to finite, quantities.

That's where the matter stood for the next two centuries.

The first person to seriously re-examine the question was George Cantor, a German mathematician who was active in the late 1800s and founded what's now called set theory. For Cantor a "set" is just a collection of things. *Any* things. You can have a set of numbers, like the natural numbers or the perfect squares. You can have a set of letters. You can even have a set of other sets. Or you can mix several types in one set; it doesn't really matter.

Sets can come in any size, but set theory only really gets interesting if you study infinite sets. Cantor took a closer look at Galileo's paradox, and built it into a much larger theory of infinity. He boldly pronounced that if you can put two sets in one-to-one correspondence, like Galileo did with the natural numbers and the perfect squares, then they are the same size. That's just how he defined "equal size"; it generalizes the normal notion of equality that we're all used to from finite numbers. Using this definition, Galileo's paradox becomes just a counterintuitive property of infinite sets. You can have one infinite set (like the perfect squares) that is a strict subset of another infinite set (like all natural numbers) and still have the sets be of equal size. That's totally legal. This property of infinite sets is a bit trippy, but it is logically self-consistent, and at some point you just have to accept it.

Another example of this property is comparing all the decimal numbers from 0.0 to 1.0 with all the numbers from 0.0 to 2.0. You can have a one-to-one mapping between the two sets, defined by doubling a number in the first set or halving a number in the second set. And yet, the first set is an interval half as large as the second set.

Infinite sets like the natural numbers or the perfect squares are called "countably" infinite. This means you can order them such that there is a first element, a second element, a third element, and so on, covering everything in the set. To be a little more precise, we say that N is the set of natural numbers (one, two, three, etc). A set S is defined to be "countably infinite" if you can put its elements in one-to-one correspondence with the elements of N. In Galileo's paradox, for example, the table we filled out gives us just such a one-to-one mapping between the perfect squares and N. In general there can be many other such mappings, and there can also be mappings between S and only a subset of N. However, if there is *any* one-to-one mapping between S and N, then S is countably infinite.

The big shocker from set theory - the reason that it is a whole branch of math rather than just a creative definition of the word "equal" - is that there are sets that are uncountably infinite. They are bigger than N; *any* mapping from N to the larger set must leave out some of its elements. In fact, there is an infinite succession of these "uncountable" infinities, each more infinite than the ones before it.

The idea of uncountable sets may seem pretty bizarre, but they are actually more common than you might think. I already brought up the simplest one: the set of all decimal numbers between 0.0 and 1.0. This set is sometimes called C, for "continuum". In a minute I'll show you the technical proof that C is larger than N. I promise it will be the only proof in this chapter! But first I want to argue it's at least plausible. Between any two elements in N, say 12 and 15, there is a finite collection of numbers. But between any two elements of C, like 0.52 and 0.53, there are infinitely many numbers. Not only that, but those numbers can have infinitely long decimal expansions. In some sense C is an "infinitely dense" set, whereas N is sparse.

Now for the actual proof. Even though set theory is generally considered an advanced subject, Cantor's proof is simple enough that anybody can follow it. He used a trick called the "diagonalization argument", which may be the niftiest math proof ever devised by humans. A friend of mine in college - a physics major who had just realized that his actual passion was theoretical math - once told me that he thought the meaning of life was contained in the diagonalization argument. No joke. For now let's just say I have my reservations about that, but Cantor's proof is indeed really clever.

Cantor did what's called a proof by contradiction: you assume the opposite of what you want to prove, and show that it leads to a logical contradiction. So Cantor imagined that N (the natural numbers) and C (the continuum) were the same size, i.e. there is some one-to-one mapping between them. Let's called that mapping F. So F(1) is the first element of C, F(2) is the second element, F(3) is the third element, and so on. The table for the one-to-one correspondence starts off like this:

Element of N	Element of C
1	F(1)
2	F(2)
3	F(3)
4	F(4)

It's obvious from the way we've laid out the table that every number in N will be present in the left column, and get a corresponding number in C. But in order for this to be a one-to-one mapping we need to also make sure that every number in C appears somewhere in the right column. For any decimal number x between 0.0 and 1.0 there has to be a corresponding integer n, such that x=F(n) and x is in the nth row in our table. It's not obvious that such an n always exists. The diagonalization argument will show that, no matter what F is, there is some element x in C that gets left out. If there is *always* an x that gets left out then we must conclude that C is more infinite than N.

The key idea for the proof is to look at elements of C in terms of their decimal expansion. Let's say that F(1) is equal to 0.23235760..., with the decimal expansion possibly going on forever. In this case the first digit of F(1) will be 2, the second digit 3, the third digit will be 2 again, and so on. We can lay out all of the decimal expansions, for all of C, in a big table like this:

	1st Digit	2nd Digit	3rd Digit	4th Digit
F(1)	2	3	2	3
F(2)	4	4	4	4
F(3)	1	2	9	8
F(4)	6	7	8	9

	1st Digit	2nd Digit	3rd Digit	4th Digit
x	3	3	2	3

These decimal expansions might go on forever, or they might terminate. If they terminate, just make all of the later entries in our table 0 for that row.

Here comes the contradiction: we will construct a number x that is manifestly in C, but that doesn't appear in any row of the table. Again we think of x in terms of its decimal expansion, and define it in the following way:
- the first digit of x is some digit different from the first digit of F(1). Maybe it's 3, since F(1) starts with 2 in our example.
- the second digit of x is different from the second digit of F(2). Maybe 5, since 5 is different from 4.
- the third digit of x is different from the third digit of F(3)
- and so on...

Defined this way, we know that x has a different first digit from F(1), so x is not equal to F(1). It also has a different second digit from F(2), so x isn't equal to F(2), and so on. In general for any natural number n, we know that x differs from F(n) in the nth decimal place. So we never have F(n)=x, and hence x does not appear in our table.

That's the key idea of the proof: for any mapping F, construct x so its nth digit is always different from F(n)'s nth digit. Strictly speaking there is still one loose end to tie up, namely how we deal with an infinitely repeating sequence of nines. For example, 0.212499999999... In this case, you recognize that this is just a perverse way to write the number 0.2125, and you define your decimal expansions to always use the version that truncates.

There you have it, ladies and gentlemen: Cantor's diagonalization argument, proving that there are higher levels of infinity.

Cantor's work was a bombshell. Many mathematicians brushed it off as abstract nonsense with no bearing on anything. A few people considered him a genius. However, a number of the leading mathematicians of the day lashed out violently against the notion of higher infinities. They thought the concept was not just laughable, but dangerous. It was a heresy, which threatened to lure young mathematicians into a career of chasing their tails. Perhaps worse, it would make the mathematical community a laughing stock. Henri Poincare, arguably the top mathematician of the age, said that set theory was a "grave disease". Leopold Kronecker, another luminary, went further and attacked Cantor personally, labeling him a "scientific charlatan" and even a "corrupter of youth". Kronecker then embarked on a personal crusade against set theory and its inventor, and succeeded in barring Cantor from employment at the most prestigious universities of the day.

Before you potentially rush to judgement against Kronecker, I should note that Cantor didn't do much to help his case. Instead he continued Pythagoras' tradition of blending math with religious mysticism. He didn't think of higher infinities as mathematical curiosities, but insights into the transcendent aspects of the world. He felt that there was some relationship between the divine and an absolute infinity which transcends all infinite numbers. It has even been suggested that Cantor felt set theory had been divinely revealed to him, rather than discovered.

Naturally, the fight spilled over from the academic community into heated debates on theology. In a different age Cantor would almost certainly have been executed for heresy, but as it is the dispute was confined to a war of paper. Christian theologians had long regarded G-d as an "infinite" being (whatever that means), and many of them saw Cantor's views as a challenge to divine supremacy. Cardinal Johannes Franzelin went so far as to equate set theory with pantheism, and Cantor spent a great deal of time defending his work in philosophical circles.

The criticism Cantor faced took a terrible toll on the man. In 1884, at the age of 39, he was hospitalized with his first bout of crippling depression, which would plague him for the rest of his life. His energy for mathematical work, his true passion, was largely sapped, and he began branching out into philosophy and literature. Despite a growing acceptance of his work, and even a personal (though not philosophical) reconciliation with Kronecker, Cantor's self-confidence, energy and sanity degraded over the ensuing decades. He spent his last years in poverty, and died in an asylum. Whatever you think of higher infinities, Cantor was a creative genius and a good man, and that was a lousy way for him to go.

Before I get into my own opinions about set theory, I should explain it's place in the mathematical community today. The terminology is absolutely ubiquitous. Almost all mathematical concepts can be described as the set of all something-or-others that have some special property. For example, a circle is the set of all points in the plane that are the correct distance away from the circle's center. When it comes time for proving theorems and hammering out the details of a subject, set theory is the most practical way to say what you mean with exacting precision.

The headier parts of set theory are still taught and actively researched, but they have become a very focused sub-specialty that is closely tied to the theory of formal logic. The community that studies it has become a bit intellectually insular, and their net impact on the world is just the massive amounts of coffee they consume. In the mathematical community as a whole people have differing opinions about the philosophical implications of set theory, but they have generally figured out by now that it's not a topic for polite conversation.

Personally, I side with Poincare and Kronecker. I've joked in the past that my contribution to set theory is the following:
> Theorem : set theory will never be useful for anything
> Proof : since humans can only do things one at a time, if we ever encounter an uncountable set we won't even be able to recognize it as such.

In all seriousness though, I think Cantor fell into the same trap that Plato did two millennia beforehand. Plato put language up on such a pedestal that he thought anything you could describe with language has to exist in the real world. Since a lot of what we can describe seemingly does not exist in the real world, he postulated the existence of the "Realm of Pure Forms", where words had a concrete reality that was even more real than the world we live in.
That's nonsense.

In the same way, Cantor didn't discover that uncountable infinities actually exist. What he really showed is that mathematical language can talk about them in a coherent way, and what that dialogue would look like. That's a fascinating thing to know, but it doesn't tell us anything about the real world. You can also talk about hobbits and wizards in a logically consistent way.

Plato and Cantor's problem is an insistence that the nature of world must reflect our descriptions of the world. But this gets the whole exercise inverted. The universe is under no obligation to conform to our descriptions of it, even if those descriptions involve something as devilishly clever as the diagonalization argument. Exploring the reaches of what language (including mathematical language) can and can't describe is fascinating stuff, but at the end of the day it's only an exercise in linguistics.

If set theory is fundamentally just a linguistic exercise, then maybe it shouldn't come as a surprise that it's actually riddled with conceptual holes. "Naive set theory", as Cantor's work is now termed, is filled with logical inconsistencies, whose existence threatened the whole endeavor. The most famous of these is Russell's paradox, discovered by the English polymath Bertrand Russell.

Bertrand Russell was one of those people who is really, really hard to give a label to. His job title was professor of philosophy, and he made contributions to basically all types of philosophy except aesthetics. His main focus though was on the blurry area between philosophy and math (though not in the woo-woo Pythagorean way) so you could also make a good case for him being a mathematician. On the side, Russell was an avid historian, social critic and political activist, and had a distinguished mini-career in each of these areas. Oh, and he won the Nobel Prize in Literature. So call him whatever you want, but he certainly deserves a place in our discussion.

Russell discovered that it was possible, using Cantor's set theory, to define a perfectly reasonable-sounding set that is a contradiction in terms. Let's call the set in question R, for Bertrand Russell. Now remember that the elements of a set can be anything - numbers, geometric shapes, or other sets. It is even possible for a set to contain itself. This idea might seem a little weird at first: a set being an element of itself, and hence an element of an element of itself, and so on? That seems a little perverse. But on the other hand, we've already given the green light to uncountably infinite sets, so let's not pass judgement.

What is Math?

Russell's set R is defined to contain precisely those sets which do *not* contain themselves. So R will contain the set N of natural numbers, since all the elements of N are numbers rather than sets. R will also contain the continuum C. Really R will contain most sets we ever talk about; self-containing sets are legal, but they're rarely interesting or useful.

Now hold on to your seats, because here is the critical question: does R contain itself? Time for a proof by contradiction again. Let's say for a moment that, yes, R is a member of R. Well in order to be in itself, then R must satisfy the criterion for membership: R cannot contain itself. Whoops! Ok, that's a contradiction. So then R is *not* in itself, right? Well not so fast. If R is not in itself, then it qualifies for inclusion in R. So all told we see that if R is an element of itself, then it must not be an element of itself, and vice versa. But it has to be one or the other, doesn't it? Well no, there is a third option. Everything we've been saying is meaningless because the set R does not, and indeed cannot, exist. It's a grammatically correct nonentity.

By the time this was discovered set theory had, despite the best efforts of Kronecker, become entrenched. Countless mathematicians were jumping on the bandwagon. Researchers before Cantor had been restricted to studying formulas and shapes that they could describe explicitly. For example, if you want to study a triangle you might break it down into the line segments on its sides and write down formulas for them. But what if you want to describe some sort of weird, amorphous blob shape? What if you want to prove a theorem not just about a well-known class of shapes like triangles, but *all* shapes, including the amorphous blobs? Set theory allowed mathematicians to talk in a coherent way about these unconventional cases, since any shape can be described as just a set of points in the plane. In fact, *every* area of math can be re-cast into the terminology of set theory if one so desires. This dual promise of greater generality and firmer foundations won over a generation of researchers.

So it came as quite a shock when Russell showed that set theory leads to nonsense.

Math isn't the only place you find logical paradoxes. According to legend the ancient Greek seer Epimenides, who grew up in Crete, once remarked that "all Cretians are liars!" Was he lying there? Even in the Bible, King David declares in Psalm 116 that "every man is a liar!" These are historical examples of the so-called Liar's Paradox. If you strip out the context of who is talking, and eliminate the nagging doubt that a liar might be telling the truth this one time, the logical paradox boils down to the following: "this statement is false". There's no getting around that one.

Natural language is filled with grammatically-correct nonentities like this. These statements are neither true nor false, but inherently meaningless. It's arguably a deficiency of language that we can even pronounce them! These things are sometimes good for a laugh, or maybe for some linguistic probing, but other than that people ignore them. They have nothing to do with the world we live in; they're just a quirk of the language.

The one area where I've ever seen people give any serious attention to logical paradoxes in natural language is in questions of religion. For example, can an omnipotent deity make a rock so heavy that He can't lift it? The Christian theologian C.S. Lewis (who is better known as the author of the Chronicles of Narnia books) summed up the prevailing opinion:

> His Omnipotence means power to do all that is intrinsically possible, not to do the intrinsically impossible. You may attribute miracles to Him, but not nonsense. There is no limit to His power.

> If you choose to say, 'God can give a creature free will and at the same time withhold free will from it,' you have not succeeded in saying anything about God: meaningless combinations of words do not suddenly acquire meaning simply because we prefix to them the two other words, 'God can.'

> It remains true that all things are possible with God: the intrinsic impossibilities are not things but nonentities. It is no more possible for God than for the weakest of His creatures to carry out both of two mutually exclusive alternatives; not because His power meets an obstacle, but because nonsense remains nonsense even when we talk it about God.

If you agree with me that math is just language, and proofs are just extensions of logical deduction, then Russell's paradox isn't problematic. The possibility of paradoxes seems to be inherent to language, so we probably should have expected something like this to be lurking in the woodwork. In practical terms Russell's paradox doesn't interfere with doing arithmetic any more than the Liar's Paradox keeps us from constructing legal arguments, so why sweat it?

If, however, you are on a borderline-religious crusade to reduce all of math to set theory - as much of the mathematical community was - then Russell's set is an inescapable crack in the foundation.

The would-be knight in shining armor was David Hilbert. One of the most prolific mathematicians of the age, Hilbert was also the founder of the formalist school of math (though it's not clear how much of a formalist he himself was). Hilbert was dismayed that something seemingly sensible like Russell's set could be inconsistent, so he set out on a quest to put all of known math, and all possible future math, on a solid foundation.

This is of course the same goal that Euclid had more than two millennia earlier, but Euclid hadn't gone far enough. In fact, Hilbert was able to go back and find places in The Elements where even Euclid had used unstated assumptions in his proofs. It's not that these assumptions were wrong. Quite to the contrary, they were so obvious that Euclid hadn't even realized he was making them! As an example, back in chapter 4 I gave Euclid's construction of an equilateral triangle, which I'll reproduce here:

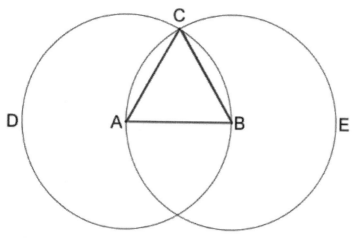

In the proof Euclid started out with points A and B. Then he drew a circle around A that passed through B, and vice versa. Then he picked a point where those circles intersect and called it C, and drew a triangle between A, B and C. Euclid's proof relies on the fact that the two circles intersect with each other, but he never actually proved that. They do of course intersect, as is glaringly obvious from the picture, but making something "obvious" is not the same as providing a proof from first principles. The fact that a slip like that could fly under the radar for two millennia spooked Hilbert. Could an oversight like it be at the heart of Russell's Paradox, and could there be other demons lurking in the corpus of mathematics?

Hilbert's solution to this problem was not just to axiomatize mathematics, as Euclid had done, but to strip mathematical statements of their meaning (at least for purposes of doing your proofs). He wanted to view mathematical statements just as grammatically-correct strings of symbols, not attaching any particular interpretation to them, so that nobody would be able to allow those pesky, fallible intuitions into the mix. Obviously we pick symbols like "=" and "1", which correspond to mathematical ideas, but for purposes of a proof they're just strings of symbols. No intuitions, pictures or understanding allowed!

The principles of deductive reasoning then become pattern-matching rules for how you can combine existing strings of symbols (like axioms) into other strings of symbols (like theorems). A string of symbols is "true" if it can be derived in this way from the axioms, and "false" if its opposite can be derived.

The process of checking or constructing a proof stops being a creative endeavor, or even an exercise in stale deduction. It is just a rote game that could be played by a diligent chimpanzee (or by a computer - more on that later). If the mathematical sentences don't have any meaning, it should pretty hard to accidentally use an "obvious" assumption, right?

This may remind you of the context-free grammars we discussed in chapter 2. In that case, CFGs are an attempt to distill a language's syntax into meaningless symbols so that we can study the grammar in isolation, and figure out how the brain parses it. It's the same idea here, but now we're focused less on syntax and more on the process of deductive reasoning itself. The technical term for what Hilbert was trying to create is a "formal system". A formal system consists of
1) a finite alphabet of symbols
2) a syntax for determining which strings of symbols are grammatically correct (usually this will be in the form of a CFG)
3) a finite set of axioms, i.e. strings of symbols we christen as "true"
4) a finite set of rules that let us combine one or more "true" statements to create another "true" one.

The so-called "Hilbert program" was a massive effort in the beginning of the twentieth century to reduce all of mathematics to a single formal system.

Hilbert knew that the human brain doesn't work as a formal system, and that day-to-day mathematics would continue to be done the old-fashioned way. His goal wasn't to change the way mathematicians actually worked or solved problems, but to guarantee that there were no more Russell's Paradoxes that could cast the discipline into doubt. Here is the real genius of Hilbert's program, and the reason he wanted to reduce things down to a formal system. "Mathematics" isn't a precisely defined term. It's a complicated, multi-faceted human activity. A formal system though, like a CFG, is perfectly well-defined. A formal system is so precisely defined, in fact, that maybe we can prove some theorems about it...

Here then is the essence of Hilbert's program, the holy grail he was pursuing. By reducing math to a formal system, he wanted to prove theorems not about numbers or shapes, but about mathematics itself. The goals of Hilbert's program were:
1. Create a formal system that captures math
2. Prove that any true statement in that system can be derived with that system, and any false one refuted
3. Prove that the system can't have a contradiction like Russell's set

One thing to make clear here is that Hilbert's Program was purely an academic exercise. Even if math could be reduced to a formal system it would be realistically impossible to use it in solving real problems. Bertrand Russell himself wrote a book about taking a purely formal approach to arithmetic, and it took him a full page to prove that 1+1=2. In a hundred years this stuff would end up becoming useful when computers were invented, but in Hilbert's time there was no reason to anticipate any applications. This was the most ivory tower of mathematics.

The sterile formality of Hilbert's mathematics stands in contrast to the warm, admirable man himself. Outside of his personal brilliance he was a mentor and loyal friend, and the generosity of his personality was no-doubt part of how he became such a leader in the mathematical community. He was also extremely egalitarian in an era where racism and sexism were rampant. He lived and worked in Nazi Germany, but despised the regime. When the Nazi minister of education, Bernhard Rust, asked Hilbert "How is mathematics in Göttingen now that it has been freed of the Jewish influence?" Hilbert replied, "Mathematics in Göttingen? There is really none any more."

He also stood up valiantly for the mathematical physicist Emmy Noether. Noether wasn't just good; she was one of the greatest minds in the history of physics. Among her many discoveries was Noether's Theorem, which is widely considered the most elegant theorem in all of mathematics. Also one of the most abstract theorems you'll ever run into, it connects conservation laws in physical systems to the underlying symmetry of the system's laws. For example, it shows that the law of conservation of energy is a consequence of the fact that the laws of physics are constant over time. Despite her groundbreaking work, Noether was constantly denied faculty positions, teaching privileges, and even salary because she was a woman. Hilbert was one of the few people who stood up for Noether as soon as her talents became evident, once famously declaring "I do not see that the sex of the candidate is an argument against her admission as *privatdozent*. After all, we are a university, not a bath house."

Hilbert had good reasons for trying to turn math into a formal system. He wanted mathematical certainty that math was self-consistent, rather than just "it looks good to me". He also wanted a way to guarantee that human intuition wasn't polluting mathematical proofs. The formal system approach was a means to these ends, a way to get mathematical certainty about mathematics itself.

However, the formalist school of philosophy took this idea and ran with it. For them formal systems were the real heart of mathematics, and human intuitions were just a useful (but fallible) guide to how the formal systems play out. The formalist school came to dominate the mathematical community, possibly because it was an excellent cop-out for the tough philosophical questions. Does infinity exist? Beats me. Is the Pythagorean theorem true? Well, it's provable if you have the right axioms. What does all this mean, anyway? Nothing.

To this day if you get a few beers into a mathematician and start asking them about philosophy, they will usually start off as a Platonist. They will talk in vague terms about how math studies timeless properties of beautiful, abstract entities which exist in a transcendent blah blah blah... But if you start to push them on where or how these "entities" exist, they will eventually back into the corner of formalism and wash their hands of the whole issue.

To my mind formalism is not just incomplete in its perspective. I also think it's a little insidious. Hilbert left the mathematical community with a culture that focused myopically on formalism and rigor, to the detriment of insight and originality. A large part of the efforts of the mathematical community since Hilbert's time have not been aimed at new and exciting problems. They have been focused on making existing mathematics more rigorous, more formal, and more abstract. Basically they have re-written the works of previous generations of mathematicians, plugging up logical holes and accounting for more edge cases, but they haven't added many new chapters. It is the view of many, myself included, that most of the interesting, exciting new math today is being done under the labels of computer science, engineering, economics and biology. Sadly, mathematicians are largely just feeding on the crumbs that are left over after scientists and engineers have devoured the cool new problems.

I think the issue is partly that humans are intellectually lazy by nature. Left to our own devices we will chew our cud on problems that we already know and understand. We are scared to venture into truly new ground for fear that our efforts will be futile or our results will end up being useless. Much better to make incremental improvements to existing ideas and settle for a comfortable mediocrity. Breaking this cycle requires a steady stream of new problems that are forced on us by external realities, rather than ones that we have chosen ourselves.

The most toxic side effect of formalism was during the 1950s, when it percolated outside of academia and into elementary education, in the form of the now infamous New Math. After the Soviet Union launched the Sputnik satellite, America flew into a panic over the idea that the Russians were training better engineers. The solution they hit upon was to give students a firm grounding in the "fundamentals", starting as young as possible.
  Children were taught about set theory before they could count, axioms before they knew how to think logically. The idea was to

build a firm foundation for more abstract concepts, but it was a complete failure. Teachers didn't really understand the material. Parents were infuriated when they couldn't help their third-graders with their homework. And the children, while they could wax philosophical about the number system an alien civilization might invent, couldn't do arithmetic. The problem was that the human brain simply doesn't process math in a formalist way. The real foundation for a mathematics education is our number sense, our notion of 3d space, and the ability to think clearly. Those can only be made rigorous after they're understood intuitively.

Ok, now that I've gotten that off my chest I need to give one very big qualifier, which I've hinted at already. The human brain may not do math as a rote game of symbols, but computers most certainly do. In recent years, the computer industry has breathed new life into formalism. While most math research is still done the old-fashioned way, automated proofs have opened up whole new areas of application. For example, if you're designing a million-transistor electronic chip, it's awfully nice to have mathematical certainty that there's no input that will make the chip freeze up.
Such a proof will generally be much too tedious for a human to work with - tens of thousands of pages of the most stale logic imaginable. But computers are fine with such tedium. In situations like this, humans work on translating real-world problems into formal systems, and then let computers do the grunt work of the actual proof.

Hilbert's program may have had far-reaching implications for how modern math is thought about and done, but in fact the program itself failed spectacularly. It's not that people couldn't construct good formal systems - they constructed a lot of really ingenious ones. The problem was a lot more fundamental than that. Possibly the most dramatic event of 20th century math happened in 1930, when a timid, gangly young logician named Kurt Godel showed that Hilbert's entire program was fundamentally in vain.

Remember that Hilbert wanted a formal system that satisfied three criteria:
1. It could capture all known math
2. It was "complete", meaning that any true statement could be proven and any false statement refuted
3. It was self-consistent

Godel proved that you can't have your cake and eat it too. Any formal system that is both self-consistent (criteria 3) and powerful enough to capture even basic arithmetic (criteria 1) will not meet the second criterion. So remember all of those clever formal systems people were coming up with? Every single one of them had true statements that they could not prove.

I promised that the idea of self-reference would crop up again in our story, and here it is. These unprovable truths are called "Godel sentences", and they are self-referential. Godel discovered a way to encode statements about formal systems as statements about basic arithmetic. The Godel sentence for a particular system is the arithmetic equivalent of "the formal system in question cannot prove this Godel sentence". The pinnacle of self-reference, a Godel sentence refers both to itself and to the formal system that it's written in. If the Godel sentence were false, it would mean that the Godel sentence is provable, which would mean the Godel sentence has to be true. Oops, that's a contradiction! So we know the Godel sentence isn't false. If it's true on the other hand, then that means the Godel sentence can't be formally proven. So here we have a statement that is undeniably true, but that cannot be proved by the formal system. Rest in peace, Hilbert's Program.

For Godel this so-called "incompleteness theorem" wasn't any kind of philosophical problem, since he was a Platonist. The Godel sentences are perfectly true, just as much as any other fact. The problem is with formal systems, and really with the whole Formalist school of thought. Godel proved that formalism is only an imperfect tool for studying mathematics; it can't all be reduced to rote, mechanized symbols. On some level that's a beautiful, humanizing result.

What is Math?

The mind of Kurt Godel was brilliant but delicate. He worked as a lecturer in Austria during the rise of Hitler, and the spectre of Nazi violence led him to have some form of a nervous breakdown, landing him in a sanitorium for several months. After fleeing the country, he almost botched his U.S. citizenship exam by fretting over inconsistencies in the Constitution. He was only saved by a very understanding judge and the intervention of his friend Albert Einstein. Toward the end of his life Godel's neuroses got worse, and he developed an obsessive fear of being poisoned. Eventually he would only eat food prepared for him by his (apparently very patient) wife Adele. When she was hospitalized for six months and unable to prepare his meals, he starved to death.

A professor of mine once said that, for mathematicians, going crazy is an occupational hazard. Perhaps he was right.

Self-reference has popped up a number of times here: in Russell's set, the Liar's Paradox and finally Godel sentences. Really, it seems to lie at the heart of most of these tricky philosophical questions.

The Liar's Paradox and Godel sentences refer to themselves blatantly, but Russell's set is also indirectly self-referential. In fact, the lack of obvious self-reference is probably why it seems reasonable at first. It's still there though: when we say that Russell's set R contains *any* set that doesn't contain itself, we are implicitly applying the selection criteria to R itself.

Indirect self-reference even shows up in the theological conundrum of "can an omnipotent god make a rock so heavy that He can't lift it?" In this case "omnipotent" means "able to do anything". If we count "making a rock so heavy that He can't lift it" as a thing, then the statement becomes self-referential and self-contradictory.

If you recall from chapter 2, self-referential language constructs are a critical feature of natural language and especially mathematical language. They may even be the definitive trait of human communication.

Self-reference is a subject with deep roots in the human psyche. Many artists have featured self-reference in their works, most famously M.C. Escher. The ouroboros, the symbol of a serpent eating its own tail, is the most classic example. It features in the Book of the Netherworld, a funerary text from ancient Egypt. The Aztec snake god Quetzalcoatl is often portrayed as an ouroboros. In ancient India it was associated with kundalini energy, a mystical energy which supposedly lies coiled at the base of a person's spine.

Some form of self-reference may even be at the root of human consciousness. After all, aren't you conscious of yourself? Our society even uses "self-awareness" as a synonym for the phenomenon of consciousness. Nobody is exactly sure what this means, since nobody really understands consciousness, but the terminology is tantalizing.

Some have even gone so far as to suggest that, because you can recognize that a Godel sentence is true even when the formal system can't prove it, your conscious mind cannot be fully captured, even in principle, by a formal system. If you wrote down a formal system describing how your mind works then that system would have its own Godel sentence. However, you already recognize that Godel sentences are true, so the formal system for your mind is able to prove its own Godel sentence.

People much smarter than myself have used this argument to say that the human mind could never be fully mimicked by computer. Maybe some parts of the brain can be simulated by computers, but "insight" and "consciousness" operate on some completely different principles that science has yet to understand. It's a fascinating possibility. Then again, other people much smarter than myself disagree. I won't opine on this subject; it's above my pay grade.

There is a final twist in the tale, one last little bit of weirdness in this most bizarre of chapters. You've already seen that there are two different sizes of infinity. The countable infinity of the natural numbers N, and the "uncountable" infinity of the continuum C. Cantor showed that there were other infinities greater than the continuum, but he never managed to find any infinities of intermediate size between N and C. After a long search, Cantor hypothesized that there were no such infinities: the continuum is the smallest uncountable infinity. This has become known as the Continuum Hypothesis (CH).

Mathematicians tried for years to prove or disprove the CH (using an axiomatized version of set theory that doesn't suffer from Russell's paradox, for the record), but to no avail. Then in 1966, a guy named Paul Cohen showed them why. It turns out that the CH is independent of the standard axioms of set theory: it can neither be proven nor disproven. It's a similar situation to the parallel postulate of geometry, which is independent of the other axioms.

For Platonists this wasn't a problem. It just meant that the in-vogue axioms of set theory (called the Zermelo-Frankel axioms) didn't fully capture the subject. The CH is still either true or false; we just need a new way to determine which. For formalists, this was a startling result... but this is all still meaningless, right? No need to get worked up about it. The intuitionists didn't care, having checked out of the argument a long time ago.

Most mathematicians consider the matter to be closed. The whole matter really - not just the CH, but all of this "foundations of math" business. It was interesting while it lasted, and it led to a lot of cool, accidental results. The original questions though, the ones that motivated a generation of mathematicians, were a failure.
Most mathematicians today are closet Platonists, and formalists if you pin them down about it, but pretty much everybody agrees that it's not longer a subject for polite conversation.

In the end it turns out that math doesn't have a single firm foundation. It's more of a mobile home - at least if you insist that the foundation be other math. The real foundation for all of this, the place where it all came from, is the human brain.

The soul-searching in this chapter doesn't tell us anything fundamental about the world we live in or the nature of truth. However, it might tell us something fundamental about the way we understand and interact with the world. Are humans forever locked in to seeing the world through linguistic eyes?

To take one semi-concrete example, theories of physics are always written down as equations, i.e. a formal system. Since the physical world is certainly complicated, and hopefully it's self-consistent, might Godel's theorem put constraints on how many questions can be answered using our laws of physics? The nature of consciousness is another tough nut to crack. Can a mind understand its own workings?

Language in all its forms - from gossip, to axiomatic set theory, to pillow talk - is humanity's great hammer. But that doesn't mean everything in the universe is a nail.

7 CONVERSATIONS WITH SILICON

"Hello, I am Siri." Those words marked a new age in human-computer interaction when they appeared on the Apple iPhone in 2011. Most computers in the past have focused on problems that are way beyond the abilities of us mortals, like calculating the trajectories of rockets, simulating the physics in a video game, or storing and retrieving terabytes of data perfectly. Siri, however, tries to do a task that turns out to be even harder: acting human.

Siri is a personal assistant application that leverages the latest computer technologies to imitate the experience of talking with a human being. She can operate other iPhone applications upon request, send text messages, make dinner reservations, and even scour the web for answers to questions. We don't have robot butlers yet, and there's a good chance we never will, but Siri is the closest the humanity has ever come to inventing a companion for itself.

So why am I writing about computers in a book that's ostensibly about math? I'll admit that this chapter is a bit of an aside, but computers are so deeply tied in to language and to math, especially in the modern day, that it's hard to untangle them. Under the hood computers operate with the cool precision of mathematical calculations. At some level of granularity there is no nuance, and the machines are perfectly digital. At a high level though they strive to cope with the subtlety and ambiguity that characterizes human interactions. This means physical layouts on a screen and touch interfaces. And in the case of Siri, it means learning to understand our language.

Perhaps surprisingly, the story of computers starts in Europe during the romantic age, with the famed poet Lord Byron and his unlikely wife. Lord Byron was a man given to excesses of every kind. He was a devoted friend, but his vicious temper was always on a hair trigger. He had numerous, passionate affairs with both women and men. He spent lavishly on just about everything. His lover Lady Caroline Lamb famously described him as "mad, bad, and dangerous to know". In his mid thirties Byron got it into his head to join the Greeks in their war for independence from the Ottoman empire, so he headed off overseas and died there at the age of 36. Before that though, Byron managed to channel all of his passion, idealism and depression into his poetry, and that's how we come to know him today.

Lord Byron had a number of children from his various affairs, but his only legitimate one was a daughter by his wife Anne. Anne was, frankly, a bizarre match for Byron. She was every bit his intellectual equal, but the similarities stopped there. He was extravagant and passionate, while she was reserved. He was a humanistic poet, while she was fascinated by math and the sciences. And perhaps most forebodingly, Byron was basically amoral (although highly idealistic, as in joining the Greek war) while Anne was strictly religious. Despite it all though, these polar opposites attracted, and they were wedded in 1814.

Things pretty much went downhill from there. Lord Byron piled up various debts, started drinking heavily, and began sleeping around. For her part, Anne became gradually convinced that Byron was going insane. Shortly after their first anniversary, and the birth of their daughter Ada, they separated.

Anne was determined that Ada would not inherent Byron's supposed insanity, so she had her tutored in mathematics and science from a young age. Among her tutors was the famous logician Augustus de Morgan, who recognized that young Ada had a real talent and told her mother she could become "an original mathematical investigator, perhaps of first-rate eminence".

Ada grew to adulthood, and her life was in many respects typical for a female aristocrat of her time. She married another aristocrat, William King, and eventually became the Countess of Lovelace (hence her more common nickname today: "Ada Lovelace"). She moved into King's mansion and settled into the role of a wife and mother, eventually producing three children. But she kept up mathematics as a hobby, and this led her into correspondence with a man named Charles Babbage.

Babbage was a prolific inventor, mathematician and philosopher. He dabbled in a number of areas, but today he's best known for designing the "analytical engine", a forerunner of modern computers. In traditional calculators the processing is all hard-coded. You can enter whatever numbers you want as input, but the logic that adds or divides them is set in stone. Babbage devised a machine where the processing logic itself could be inputted too. The device could be re-programmed at will, with instructions inserted in the form of punch cards.

Babbage became acquainted with Lovelace through social circles and was struck by her brilliance in math and writing, dubbing her the "Enchantress of Numbers". They became good friends, and began collaborating on Babbage's calculating machines.

Babbage worked on the physical design of the engine itself, while Lovelace focused on the mathematical side. The Italian mathematician Luigi Menabrea (who was also an accomplished general and statesman - lots of polymaths in this chapter) was also in the loop. He wrote a book about the analytical engine and its capabilities which was translated by Lady Lovelace from Italian to English. Lovelace included notes with her translation, which ended up ballooning to be longer than the book itself, and showed that she understood the machine's capabilities far better than any of her collaborators. Among Lovelace's various insights is the first known computer algorithm: a series of instructions for how the engine could calculate the Bernoulli numbers, which are important in certain areas of math. Sadly manufacturing issues prevented Babbage's engine from ever actually built, but in hindsight we know that, if it had been, Lovelace's algorithm would have worked perfectly.

Babbage designed the analytical engine only to solve math problems. Essentially it was a souped-up, programmable calculator. It was Lovelace who realized that Babbage was only seeing the tip of the iceberg. Depending on how you programmed the machine and interpreted the data, it could be used to do a lot more than crunch numbers. It could, for example, also process music or text instead. Really the engine was just a machine for applying logical rules to strings of 0s and 1s, so anything that could be encoded into 0s and 1s was fair game. That includes math and text of course, but that's just the beginning. Web pages, operating systems and Siri fit the bill as well. The whole information age is founded on this key insight: under the hood, everything that a computer can do reduces to the same game of symbols.

The coolest thing about computers, at least from a mathematical perspective, is that they're precise. The inputs and outputs are exact, the processing is cool and consistent, and that precision stays there no matter how complicated a computer program is.

Anybody used to modeling the real world can appreciate that luxury. Equations in engineering are only correct to within some margin of error. Then, in a vicious cycle of garbage-in-garbage-out, those errors get amplified as you proceed through a calculation. Biology is even worse, where key equations describing things like population growth are really just gussied-up heuristics. Even in physics, the most pristinely mathematical of the sciences, something like dust in your laser setup inevitably creates a gap between predictions and reality. Almost the only areas where math is actually precise areas are ones where humans artificially impose precision on the world, like accounting.

This isn't a book on the guts of computers, in the same way that it isn't a tutorial on any specific math subject, so I don't want to get into the nitty-gritty of how they work (and honestly, I'd be in over my head if I tried to get into too much detail). However, the key thing to understand is that a computer program is basically a long, complicated, precise recipe book. The basic ingredients are data. It could be numbers. It could be text. It could be an image file or a Word document, or it could be a combination of all of these. The steps in the recipe are specific to the kind of data being processed. If you're processing text you could capitalize it, append it to other text, or rearrange the letters. If it's numbers you can do arithmetic, check whether a particular number is less than 0, and so on.

To give you an example of how this looks, here is a recipe for calculating the smallest number in a list (with 0 if the list has nothing in it). We loop through the list, keeping track of the smallest number we've seen so far. By the time we get to the end of the list the smallest number so far will be the actual smallest number, so we "return" this value as the output of the recipe.

```
Input: a list L of numbers
Algorithm:
    min_so_far = 0
    For number x in L:
        if x < min_so_far:
            min_so_far = x
    Return min_so_far
```

There is more than one way to skin a cat, and you could have written many other recipes that would have given you the same answer. For example, you could have calculated the smallest number in the first half of the list, then the smallest number in the second half of the list, and then taken the smallest of those two numbers as your final output.

In computer science parlance a recipe like this is called an "algorithm". The first real computer algorithm was the one written by Ada Lovelace, to calculate Bernoulli numbers. The key thing about an algorithm is that every step is 100% precise. There is no "pinch of salt" or "add approximately 2.3 to a number". Every step is black-and-white.

Let me give you a final example, one that's a little more interesting. Let's say you have a bunch of emails, and you want to find all the people you have exchanged emails with. Not the spammers who you didn't reply to, or the people you reached out to who never wrote back. We want the people for whom the communication has gone both ways, and here is an algorithm that will find them all. It works by going through all the emails you have sent, and for each one it sees if the recipient has also sent you an email. Whenever such a person is found they are added to people_already_counted, a list of all your correspondents.

```
Input:
    SentMessages: list of emails you have sent
    ReceivedMessages: list of emails you have received
Initialize:
    PeopleYouMailed: empty list
    PeopleWhoMailedYou: empty list
    Correspondents: empty list
Algorithm:
    For email_sent in SentMessages:
        other_person = email_sent.recipient
        if other_person not in PeopleYouMailed:
            add other_person to PeopleYouMailed
        else:
            go on to next email_sent
    For email_got in ReceivedMessages:
        other_person = email_got.recipient
        if other_person not in PeopleWhoMailedYou:
            add other_person to PeopleWhoMailedYou
        else:
            go on to next email_got
    For person in PeopleYouMailed:
        If person in PeopleWhoMailedYou:
            add person to Correspondents
    Return Correspondents
```

Designing algorithms like this is tricky. Notice, for example, that before I add a person to the list `PeopleYouMailed` I check whether they're already there. If I hadn't done that, the same person could occur multiple times, and there would be many duplicates of them in `Correspondents`. In fact, I'm embarrassed to say that you're looking at the second version of this algorithm: I made that exact mistake when I wrote it down the first time. Little "gotchas" like this are what makes computer programming so hard. The computer will do exactly what you tell it to, no more and no less. The onus is on the programmer to make sure they've crossed all their Ts and dotted all their Is.

Human programmers think of their data in terms of text, numbers, emails and other abstractions. Under the hood though, computers don't distinguish between them. It's all stored as "bits", little pieces of data that can be either a 0 or a 1. Physically, bits are stored in tiny electronic components that can be turned on or off, like little light switches. From the computer's perspective it's only job is to shuffle around and manipulate the bits according to logical rules.

Groups of bits are used by humans to encode more complex data. For example, a letter of the alphabet is usually stored as a group of eight bits, together called a "byte". The most common way to encode letters into bytes is called ASCII - the American Standard Code for Information Exchange. Here are the first several ASCII encodings:

Physical bits	letter
01000001	A
01000010	B
01000011	C
01000100	D
01000101	E

Here's the interesting thing though: computers don't know ASCII. The computer will just see the physical bits and process them according to logical rules. In fact the exact same bits are often used to encode integers in binary, as in this table:

Physical bits	integer	letter
01000001	65	A
01000010	66	B

01000011	67	C
01000100	68	D
01000101	69	E

The computer can't tell whether 01000001 refers to A or 65. Turning A into B is the same operation as adding one to 65. It's up to the humans to impose meaning on the data.

Siri may seem to speak English pretty well when she talks about restaurants and driving directions, but her native language is something called assembly code. Assembly code has no text, no numbers, no internet. It's the lowest level of computer programming, and all it describes is how the computer should shuffle bits of memory around and manipulate them. The computer programs that Ada Lovelace designed were written in something equivalent to assembly code, and the earliest electronic computers were also programmed in assembly.

In recent decades though we've developed more user-friendly interfaces. If you go to Google or Microsoft, for example, you'll see people writing code that looks very similar to the algorithms I gave you above, which talk about lists, integers, letters and the like. Abstractions that you're familiar with from everyday life. Before the algorithm gets run on a computer though, these instructions get translated into the low-level assembly code. I mean, you *can* write in the native assembly if you want, but it's pretty masochistic.

Babbage thought of the 0s and 1s in his analytical engine as a way to encode numbers. Nothing more. It was Lovelace who realized that something like ASCII was possible. She had the key insight.

The next stage in the history of computers was actually an offshoot of David Hilbert's work on the foundations of mathematics, which I covered in the last chapter. As you'll recall, Hilbert tried to make a mathematical study of mathematics itself, aiming to prove that it was free of self-contradictions like Russell's paradox. The project went down in flames in 1930, when Kurt Godel showed that any formal system has true statements that it can't prove. As much of a blow as this discovery was to Hilbert's program though, it was only one of the irons he had in the fire.

For a lot of math problems there is a straightforward (though maybe long and tedious) procedure you can follow that is guaranteed to give you the answer. Calculating the sum of two numbers is one example of this, or doing long division. It might surprise you to learn that most of high school algebra falls into this category too. People usually try to solve algebraic equations by finding a clever way to go about it, rather than just turning the crank, but most equations can also be solved purely by rote. In theory it could be done by a well-trained, diligent chimpanzee. In practice, it can be done by a computer that has been programmed with the right algorithm.

Other problems though seem to require some genuine creativity. You can try a lot of different approaches, but there's no guarantee that anything is going to work. One of the straightforward approaches might pan out, but your blood and sweat may also have to be supplemented by a "eureka" moment. In the end it could turn out that there is no solution, or maybe that you didn't manage to hit upon the key idea. Most of mathematical research falls into this category.

Hilbert was fascinated by this dichotomy. On the one hand some problems can be solved purely by rote, or what he called a "mechanical procedure" (computers didn't exist at the time). A mechanical procedure can involve some guess-and-check, but the guess-and-check has to be all specified ahead-of-time and written into the algorithm. On the other hand some problems *seem* like they require some spark of insight. Or at least, nobody to date had figured out a mechanical procedure for solving them.

So Hilbert posed a grand challenge to the mathematical community, sometimes called the "Entscheidungsproblem": is there a mechanical procedure that, at least in principle, can either solve any math problem, or determine that there is no solution? Godel showed that some true statements are not provable, but Hilbert hoped that it was at least possible to calculate whether or not a proof existed.

Let me give you an idea of how this might work. Say you start off with a formal system describing something like arithmetic. The system will include a finite number of axioms and a finite number of inference rules. Those axioms and rules might include something like this:
1. Axiom: 1=1
2. Rule: if x+1=y, then x+1=y+1

You can then use these to start chugging out provable statements. You have 1=1 as a given, so then you can prove in one step, using the inference rule, that 1+1=1+1 (i.e. 2=2). Given this result, you can apply the inference rule another time to prove that 1+1+1=1+1+1. You can keep going on and on, chugging out true statements ad infinitum. Obviously this is the worst kind of grunt work, suitable only for a computer, and perhaps even beneath a computer's dignity. But at least in principle you can program a computer to chug out all provable statements for a formal system.

The computer will be spitting out theorems forever, but everything that actually has a proof will be proven eventually. It takes some care to make sure that you don't just go off on a tangent, like applying a single rule forever and never actually getting to some other the other, but it can be done.

So here's the idea. Let's say you have a statement S, and you want to know whether it can be proven. So you set your theorem-proving computer to work, chugging out all possible proofs for your formal system. If S is actually provable then, eventually, the computer will spit out a proof of it and we can declare victory. Conversely, if S is dis-provable then eventually the computer will spit out a refutation. In either case we are guaranteed to get our answer. But what if S falls into that gray area, of statements that can't be proven or disproven? In that case our theorem-prover will just keep running forever. It will never give us a proof of S, or a refutation, and we will never get a definitive answer. Hilbert hoped (and it looks like he was pretty optimistic) that this tangle could be ironed out, and that it was possible to figure out ahead of time whether the theorem prover would give us an answer about S.

Alas it was not to be. In 1936 - six years after Godel - the Englishman Alan Turing showed that even this consolation prize was unavailable. There was a huge class of math problems that in principle couldn't be solved by a computer, and this was among them. Special cases could perhaps be solved, and those solvable special cases might even cover most of the problems in real life, but there was no way to program a computer so that it covered every possible scenario.

Ada Lovelace is the one who invented the idea of algorithms, and realized how powerful they could be. Turing is the one who began to study them in a systematic way, and in so doing he laid the theoretical groundwork for modern computer science. His key contribution was the so-called "Turing machine", an abstraction meant to precisely define what a "mechanical procedure" actually entailed.

That really was the crux of the issue. Hilbert understood intuitively what he meant by a "mechanical procedure", and that's the terminology he used when he issued his challenge. But if you want to prove any theorems about what "mechanical procedures" can or cannot do, you need to flesh the idea out into a rigorous definition. In the same way that Hilbert had sterilized mathematical deduction down into formal systems, we need some distillation of what an algorithm is.

A Turing Machine is an abstract hypothetical machine, sort of an idealization of Babbage's analytical engine. As with formal systems, you have a finite list of symbols. Just "0" and "1" if you want to be true to the physical implementations of computers, but you could also use something a little more human-friendly, like letters of the alphabet. So long as there are only a finite number of them, it doesn't make a difference. The "memory" for a Turing machine is an infinite list of these symbols. Think of it as an infinite strip of movie film, where each frame contains exactly one symbol, and the symbol in a frame is potentially subject to being changed during a computation. All of the data - input to the algorithm, output, and any intermediate steps - will be stored on this tape, perhaps encoded as ASCII or something like that.

The logic of a Turing machine - the part that you can program - is a gigantic flowchart. At any given step in the computation you are only looking at one symbol in the tape of memory. You look at the current symbol, and then consult the flow chart. Based on the symbol you're looking at and the current state of the flow chart, the chart tells you

1. What symbol to replace the current one with (if any), so that you can modify the machine's memory
2. Whether you should move to the left or right on the tape to find the next symbol for the computation
3. The next state in the flow chart you should move to

One of the states in the flowchart is special. It's called HALT. If you ever end up there it means the computation is finished, and the contents of the memory at that point is the output of the computation.

Turning machines are a lot like assembly code. There are no numbers or lists, or anything familiar. It's just a cool, precise shuffling of symbols according to logical rules. Just like assembly code or Babbage's analytical engine though, you can encode higher-level abstractions like lists and looping behaviors into a Turing machine's flowchart.

Alan Turing was able to show that many well-posed problems cannot be solved by a Turing machine, and hence can't be solved by a computer. The most famous of these was the so-called "halting problem". In yet another instance of self-reference, it turns out that you cannot program a Turing machine to determine whether or not another Turing machine will ever reach its HALT state and complete its computation. Turing showed that many important mathematical problems reduced to halting problems, and they were all equally impossible.

The caveat here is that Turing just asserted that his device was the distillation of a "mechanical procedure". Was it possible that there was some other, better definition that permitted a little more power? Several alternatives were proposed by other mathematicians, most notably the American Alonzo Church, who invented an alternative called Lambda Calculus. Lambda calculus looked absolutely nothing like a Turing machine. If a Turning machine is a flowchart with some memory added on, then lambda calculus is more like the process of plugging a number into a formula, plus a weird device for plugging a formula back into itself. Personally, lambda calculus hurts my head a little, and there's a reason students usually get taught about Turning machines instead. However, in the end it turned out they are equivalent. Any problem that can be solved by Turing machines can be solved by lambda calculus, and vice versa. The fact that these two unrelated approaches ended up being completely equivalent convinced people that this was indeed the best definition of "mechanical procedure" to use, the one that most perfectly captured human intuitions.

You've gotta feel a little bit bad for Hilbert, to have had so many of his ambitions for the mathematical community dashed to pieces by young upstarts. First Godel shows that some statements are true but not provable, and then Turing shows that you can even tell which ones they are. But from the ashes of Hilbert's dreams were born the theory of computing, the discipline of mathematical logic, important advances in philosophy, and the axiomatic way that much of modern math is done. I think he would be proud.

A few years after Turing's work on computability World War II broke out. Turing joined the war effort at Bletchley park, a large, beautiful, Victorian estate nestled in the English countryside. It may have looked serene and low-key, but Bletchley housed one of the most top secret and critical wartime projects of Britain. It was here that an army of mathematicians, most prominently Turing, worked feverishly to break the Axis encryption schemes. In a highly technological war the Germans devised cipher technologies that were head and shoulders above anything from previous eras, and chief among them was the famed Enigma code. Turing and

his men used traditional cipher approaches in their work, but they supplemented human ingenuity with the raw processing power of specially-made machines that we would now call computers.
Between man and machine they were able to unlock most intercepted messages, and eventually succeeded in cracking Enigma itself. Never once, for the whole duration of the war, did Germany suspect that its invincible codes had been broken, and Winston Churchill referred to the obscure enclave of mathematicians as "the geese that laid the golden eggs - but never cackled".

Unfortunately, Turing's tremendous services during the war effort were not reciprocated by the British government after hostilities ended. Turing was gay, at a time when homosexuality was illegal throughout the United Kingdom. At one point a former lover tried to burglarize Turing, and he reported the incident to the police.
When it came out that Turing had been romantically involved with the intruder, he was convicted of gross indecency and given a choice of prison or chemical castration through estrogen injections.
Turing chose the injections, which destroyed his libido, made him impotent and gave him breasts. As a convicted homosexual, he was also stripped of his security clearance and barred from the work he had been doing on next-generation computers. In the end Turing took his own life, eating an apple he had laced with cyanide.

I would like to raise an interesting possible parallel between computers and language. This isn't a proven result or anything, but if you'll permit me I'd love to speculate a bit.

In theoretical computer science a computer language is said to be "Turing complete" if it can perform any computation than a Turing machine is capable of. So that means lambda calculus, assembly language, Babbage's analytical engine, and almost all modern programming languages are Turing complete. All of these languages have the same computational power in principle. Even though they are superficially night and day, these languages are all, in a very fundamental sense, the same.

Back in chapter two I discussed the idea of a "universal grammar" in natural languages, an idea which was pioneered by Noam Chomsky. I suspect that there is room for a notion of "Chomsky complete" languages as well - a more formal definition of what it means for a language to have the expressive power contained in the universal grammar. There are some obvious differences, such as the fact that any computer's input and output will be zeros and ones, whereas different languages have different vocabularies. But there is probably a precise way to say that different natural languages have the same basic expressive power.

I said that almost all programming languages are Turing complete, but there are exceptions to this rule. In some cases an essential element of Turing-completeness just isn't worth the trouble to implement in a real-world system. For example, languages designed for programming large clusters of computers are often not Turing complete, because of the logistical nightmare required to shuffle data between the different computers. For most languages that are not Turing complete the missing ingredient is recursion. In computer code you can have a subroutine call itself, in the same way that you can embed a complete English statement into a larger sentence, and this is often the most difficult part of the language to actually implement.

The Piraha language may be similar to these programming languages; spectacular for the applications they were designed for, but missing basic recursive functionality. If people ever come up with a satisfactory definition of "Chomsky completeness", I have two predictions about it. First, grammatical recursion will play a central role. Second, it will in some deep sense (the nature of which I can't very well speculate on) be the same thing as "Turing complete". After all, these things are all languages.

So now the big question: will Siri ever be able to think? I know any regular user can attest that, while Siri may be cool and useful, it's painfully obvious that she is just a computer. There's not really a mind in there. Might that change in the future?

First off, as you surely know, the jury is out on that question. People have been asking it ever since computers started becoming popular, and it was Turing himself who proposed a way to answer it. The first question is what we even mean by "thinking". Is it a synonym for "self-awareness", or "consciousness"? If so, what do those terms even mean?

Turing realized that trying to define "self-awareness" was a deep philosophical rabbit hole, and people would probably be unable to even agree on a good definition. So instead of defining what consciousness means, Turing designed a way to test for it. The idea is that even if we can't put our finger on what consciousness is, we know it when we see it.

Imagine you're sitting in a room at a computer, and you're involved in two text-based conversations. Maybe on Skype or Google Chat. One of your correspondents is a normal human, although it's somebody who you have never met before and don't have any rapport with. The other one is a computer program like Siri, which is programmed to mimic human conversation. Your job is to decide which one is which. A computer that can consistently convince you that it's human, even if it's only a modest fraction of the time, passes the Turing Test and, according to Turing, is able to "think".

During the test you can ask any questions you want, or bring up whatever topic you think might help you make a decision. You might start by just asking both of your correspondents point blank: are you human? They'll both answer yes of course, so that's a dead end. Then you might move on to quizzing them about daily life, or about their emotions. These are the kind of questions that are intimately familiar to humans, but Siri could be programmed with pre-canned responses to most of them. Siri can tell you that she feels lonely sometimes, or parrot lines of poetry. She can even be trained to recognize slang. And of course, she will never show off any of her skills that humans don't have, like rapid number crunching.

I should tell you right now that no computer that has ever been made can compete up to this point. If you ask probing questions about what it's like to be human, Siri will fall flat on her face (eh, screen). This failure to date isn't a fundamental limitation on Siri though, but rather a symptom of her imperfect engineers. It's an open question whether, in principle, some future version of Siri might pass the Turing Test.

So let's say a better Siri gets made, that can make small talk about daily life and has a lot of common sense coded in. Can you tell the difference now? You'll have to probe deeper, asking questions that go beyond the minutia of daily life and require original thought. The kind of questions that no engineer would have thought to hardcode. Maybe Siri can ape responses to millions of different questions, but there has to be more to a mind than that.
There is some mechanism at the core that actually understands the questions... isn't there?

In the previous chapter I mentioned briefly that some people think it's impossible for computers to think. Chief among them is the Cambridge mathematician Roger Penrose. One of his specialties is finding mathematical problems that computers can't solve. The most famous of these problems is called the Halting Problem that I mentioned above. There are many others that are less esoteric.
Penrose, for example, specializes in finding geometry problems that are impossible for a computer.

Humans like Penrose, however, seem to be able to make some headway with these uncomputable problems. We can recognize that Godel sentences are true. We can often reason out whether a Turing machine will reach HALT or not. According to Penrose, we are only able to do this because we have an "insight" into the problem that absolutely, fundamentally, can't be replicated by Siri.
No matter how much common sense you hard-code into her, Siri will never have true self-awareness.

Of course the consciousness mechanism (whatever it is) is presumably not limited to solving abstract math problems. Instead, it's the core of our self-awareness. A lot of the human brain works like a computer. Our vision and memory systems, our physical coordination: all of these tasks can be done by Siri, assuming that she's correctly programmed. She can even do them better than we can in many cases. But there is an extra little spark at the core of it all that Siri can never have. This spark is the "ghost in the machine", the place that our self-awareness comes from, and it is a necessary ingredient for the insights that let us solve uncomputable problems.

Penrose's view is a minority one. Turing himself, in fact, thought that computers would eventually be able to pass the Turing test. Most modern computer scientists agree with Turing, although none of them see it as a realistic near-term possibility. On the other hand, nobody really understands what consciousness even is, despite the fact that humans have had it for quite a long time. The question remains open, and personally I'm looking forward to seeing whatever answers may be found.

The first time I learned to program computers was in my senior year of high school. By that time I was a card-carrying math nerd and planning to become a scientist, but I was blown away by computer programming. It had the same kind of precision as math, but it seemed much more flexible and free-form. I was no longer limited to expressing things in equations. I could tell the computer to use an equation if I wanted, like in a math class. However, I could also have it run an operation only if certain conditions were met, it was easy to put in special edge cases, and the end product was a useful and interesting piece of software, rather than just another equation. Mathematics is a purely abstract exercise, but with computer science I was forced to balance that against the need for an interesting end product, the performance capabilities of the computers themselves, and the question of whether my code is clear enough that somebody else can understand it. I'm the kind of guy who likes to have his fingers in a lot of pies, and that's why I usually pay my bills doing computers programming; there are enough moving parts that it always stays interesting.

These days anybody who wants to call themselves a mathematician (at least outside of academia) needs to be conversant in the language of computers. They are the workhorses of most applications, and great things happen when you combine their brute strength with the theoretical insights of a human. I'll get more into that in the last chapter of the book.

However, it's not just mathematicians who need to understand computers. They are ubiquitous, but they are also deeply alien. Our languages and ideas are no more accessible to them than assembly language is to us. So we meet in the middle, with compromises like high-level programming languages and Siri. Perhaps this is the reason young people are famously more adept with computers than adults: they're still at the age when they absorb new languages by osmosis. Working with modern technology isn't like using a hammer or a toaster; there is so much processing power on the other end that it functions, in a lot of ways, less like a tool and more like an alien mind. It's well-intentioned, and immensely powerful, but it's slow on the uptake, even as we have trouble translating our intentions into its precise language.

Using technology is sort of like being a tourist in a foreign country, scouring our phrase book for something like "where is the bathroom" and latching on to the one person who speaks a few words of broken English. Anybody who has been in this position can appreciate how the sounds, syntax and slang of the languages often just don't mesh, but with flexibility, honest efforts and a sense of humor on both sides you can get the point across. It's the same way with technology; with enough practice and an open mind, even if we will always speak with a "human accent", we can learn the language of our silicon counterparts.

8 WHY NOBODY ACTUALLY UNDERSTANDS ECONOMICS

The title of this chapter is not news in light of the events of recent years. In 2007 the world discovered that its economic system was like a house built on a foundation of sand. Some of the most on-paper intelligent people in the world, and the governments that regulated them, worked together to create a system which hid risk not by eliminating it, but by sweeping it under the carpet. Then in the midst of what looked like a tidy house, the mess beneath the carpet exploded into the living room.

There are many, many factors that contributed to this complex problem, and I don't pretend to have the final answer. Unfettered greed and outright lying were certainly a big part of it. However, that kind of straight-up malice probably played a smaller than perverse incentives, both those which are inherent to the financial industry and those which have been aggravated by the regulatory environment. At the end of the day, the fact is that most people throughout the economy thought they were making rational decisions based on the numbers in front of them. Home buyers taking out predatory loans thought they would be able to make their payments. Institutional investors buying fractional ownership of these loans thought they were protected by diversification across a wide array of mortgages. The bankers who facilitated the process believed, at least for the most part, that they were peddling legitimate investments.

It's not that the numbers people were working with were wrong. There are always some borrowers who lie about their incomes, and some lenders who lie about their rates, but these are the minority. Numbers by themselves don't lie, but they don't tell the whole story either when they're describing real-world situations. This is where the disconnect occurred; people based their analyses and projections on the data available, but they weren't able to connect the math back to the real world, and the conclusions they drew were both wrong and dangerous. In this sense the financial crisis was a catastrophic failure, by people at all levels, to understand the real meaning of the numbers in front of their eyes.

There are many areas where math has been catastrophically misapplied. I don't just mean times when people accidentally perform the wrong calculation; anybody can do that, and such errors usually get caught before they have the opportunity to cause trouble. Obviously black-and-white errors in arithmetic or algebra need to be avoided, but they are a separate issue that is usually easy to diagnose. I'm talking about the much more insidious times where people do all the right calculations, but they lose track of the connection between math and the real world. Their math itself is done right, but people fail to include the key ingredient which is nearly impossible to fit into equations: common sense.

A "model" is the typical jargon for a mathematical description (usually an idealization) of how the world behaves. For example, a formula describing the exponential growth of a population of bacteria might tell us that ten hours from now there will be exactly 1000003.72 bacteria in the petri dish. A model for the stock market is usually random, and won't tell us exactly what the price for a particular stock will be in a year. However, it will tell us exactly how likely it is to cost more than $100. Some models, like in physics, are considered so accurate that they may as well be absolute truths. We are so confident about Newtonian mechanics that if our predictions are off we usually blame the clumsy scientist or look for an error in our math, rather than questioning the underlying laws. In other areas, like in the growth of bacteria or the movement of the stock market, everybody knows that the models are more crude and descriptive, and nobody expects to make fine-grained predictions with them. In either case though, models are mathematical descriptions of how the world operates, and the same math can be used to describe either situation.

Its illuminating that we also use the word "model" to refer to scaled-down versions of physical objects, like houses, trains and airplanes. The overall structure looks the same as the original object, except for the size. If you look more closely though, you see that many details are left out. I have a model of a Lamborghini Gallardo on my desk, which looks pretty accurate at first glance. The proportions are right, the orange color is faithful. The doors and trunk open like on a real car. You can even pop the hood. If you look closer though you see there are no locks on the doors, the steering wheel doesn't turn, and the whole thing is held together by screws. This isn't misleading to me; I know that a real car would have working locks, a steering system and good manufacturing, and I understand why those detailed mechanisms were left out.

But I only know this because I'm familiar with cars, and it's obvious to me which aspects of my model correspond to a real Gallardo, and which are specific to the toy version. If I were a hunter-gatherer who had never seen a car in my life, I might conclude from the model that real Lamborghinis say "made in China" on the bottom.

In the same way, imagine a financial manager armed with a formula for mortgage prices. It's likely that he doesn't really know how his analysts derived the formula, and his eyes might have glazed over a bit when they were telling him about all the technical reasons why it's not perfect. What he does understand, however, is that the formula is making money. Lots of it. And in fact, if he had been more aggressive with the formula over the last month, the firm would have made an extra hundred thousand dollars. See where I'm going with this? Poor understanding plus cognitive bias makes for a dangerous combination.

English descriptions of the world are like the models of trains and buildings; they are usually well within our intuitive scope, so common sense covers over a multitude of inaccuracies. There are always exceptions and corner cases, but generally everybody in the conversation is familiar enough with the topic that the disclaimers aren't necessary; we can fill in the details on our own. In fact it can be quite a chore to say *exactly* what you mean in English - that's what we pay lawyers to do when they draw up contracts.

The care-free speech of day-to-day life is a luxury that we can afford only because we are intimately, intuitively familiar with day-to-day life.

Advanced math is not like everyday English. Very few people have good intuitions for it, and those who do only get them through years of training, so it is easy for the shortcomings of mathematical models to go unnoticed. It's like a hunter-gatherer concluding from my model Gallardo that the real car has opening doors and a transparent windshield, but also a gear stick that's stuck in place. It's not that math has more failings than daily English. We're just a lot worse at spotting them.

The most infamous cautionary tale about the abuse of mathematics in economics is Long Term Capital Management (LTCM), a hedge fund which almost brought the world economy to its knees in the 1990s. LTCM was founded in 1994 by John Meriwether, the former vice-chairman of bond trading at Salomon Brothers. Something of a pied piper of finance, Meriwether brought with him many of the brightest minds in money at the time. There was Larry Hillibrand, the mathematical genius who had been the top-paid trader at Salomon Brothers. David Mullins Jr. had been the vice-chairman of the Federal Reserve before joining. Eric Rosenfeld had been a professor at Harvard before going to Wall Street. But the real feathers in Meriwether's cap were Robert Merton and Myron Scholes, who later won the Nobel Prize in Economics (technically the "Sveriges Riksbank Prize in Economic Sciences in Memory of Alfred Nobel", since Alfred Nobel didn't create a prize for economics. It was added later) for developing the Black-Scholes formula, a mathematically sophisticated way to calculate the "reasonable price" of financial options. This dream team was the toast of Wall Street, and under Meriwether's leadership they created the most celebrated fund in the world.

LTCM specialized in a trading strategy called arbitrage. In traditional investing you buy something and hold it, hoping that the price goes up over time. Buying a share of stock or a house falls into this category. This is the oldest, most common, and certainly most straightforward way to invest, but there are other strategies.

Buy-and-hold amounts to placing a bet that the price of something will go up over time. You can also bet that the price will go down, through a financial tool called a "short". To short a stock that you think will tank, you might borrow a share of the stock from me (for a nominal fee, of course) and sell it on the open market. Then later, after the price has gone down, you buy a new share at the lower price, return it to me, and pocket the difference.

Mathematically speaking, investing is all about making calculated bets of one form or another. If there is any prediction you can make about what will happen in financial markets, you can be assured that there is a way to bet on it.

Arbitrage, LTCM's main game, is a little more complicated than buy-and-hold or shorting. You don't make any predictions about whether the price of an individual security will go up or down.

Instead, you are betting that the prices of two different securities will get closer together. Imagine that security A and security B should, for some very sound financial reasons, be worth the same amount. But temporary market conditions have made it so that A is trading at $100 a share and B is trading at $110. You might not know what their actual "fair price" is - you just know it's the same thing for each security, and you want to bet that they'll eventually be trading at the same price. What you can do is buy A for $100, then borrow a share of bond B from me and sell it for $110 - you go long on the cheaper security and short on the more expensive one. The net result will be that you now have $10 and one share of A, and you owe me a share of B. When the prices eventually converge, you can swap your share of A for a share of B on the open market (you'll be able to do that since they'll be trading at the same price) and give it back to me. The extra $10 was pure profit.

In the case of LTCM, securities A and B were generally US treasury bonds. Immediately after a round of bonds is issued they are called "on the run", and there is a flurry of trading. It's easy to buy or sell an on-the-run bond, since they have a very vibrant market. Eventually those bonds settle down into pension funds and endowments, and fewer of them will change hands on any given day. At that point the bonds become harder to liquidate quickly, and hence less desirable as an asset. The price drops accordingly, and they are called "off the run". LTCM's bet, which they placed over and over again, was that on-the-run bonds eventually turn into off-the-run bonds, and at that point the prices between them will converge.

Arbitrage in its purest form is a very safe business, because it's essentially guaranteed to make money. The problem is that the margins are generally extremely low. The differences between on-the-run and off-the-run bonds, for example, are often only fractions of a percent! Sure you can make money on the transaction, but you would have made more by just putting it into a savings account. Hardly the stuff of high finance.

To magnify their returns to acceptable levels LTCM borrowed money. A lot of it. They used this borrowed money to purchase vastly more off-the-run bonds than they could actually afford, then they borrowed and sold an equally gargantuan amount of on-the-run bonds. So at any point in time they were sitting on a massive pile of assets and a massive pile of debt, slowly slowing liquidating the assets to cover the debts (plus a little extra for their own profit), as prices converged. At its peak LTCM had a trillion dollars in bonds. The deal itself was still sound, but now there was a catch.
The prices of the bonds didn't just have to converge - they had to converge fast enough that LTCM could pay the interest on all that money it borrowed. The trade itself was still based on solid economics, but LTCM's models for how fast the prices converge were based on historical precedent and sophisticated math. They bet the fund on the accuracy of those models.

It worked for a while. In fact it worked spectacularly well, with LTCM giving investors returns of 40% per year for the first four years. This fueled their confidence and boosted their reputation; armed with these mathematical oracles, they were invincible!
Tellingly, when the inefficiencies they were capitalizing began to dry up a little, they didn't scale down their operation. In fact they began to return money to their investors without reducing their

bond portfolios. This amplified their personal profits, but it did so by ballooning their ratio of debt to actual cash.

I've always been interested by the fact that the first people to sound the alarm at LTCM weren't the tried-and-tested bond traders. It was Scholes and Black, the ivory tower academics. In spending their careers putting together idealized mathematical models, including the options-pricing model that netted them the Nobel Prize, they were painfully aware of how idealized these kinds of models actually were. They saw the growing mania and hubris on the part of the traders, and knew that they were pushing the predictive power of these formulas to their breaking point.

The music stopped in August of 1998. The bond markets had been evolving recently, so that LTCM's carefully tuned formulas were already a little dated and starting to lose money. But the real disaster was the Russian Debt Crisis, where the nation of Russia simply decided not to honor its bonds. In daily life, if somebody defaults on a home mortgage or a car loan the assets can be repossessed, but there's no such recourse if a sovereign nation decides not to pay up. Everybody who had bet on the good faith of the Russian government was simply out of luck.

It's scary to get so rudely reminded that a government bond is, ultimately, based only on trust. Russia violated that trust, and it sparked a "flight to quality", where investors dumped any bond that could possibly be construed as dubious. The single most trusted debtor in the world was the US government, and investors flocked to buy up on-the-run US treasury bills… exactly the ones that LTCM was betting were overpriced. Week after week they watched as loan interest gutted their cash reserves, while price differentials remained stubbornly high, and by September the fund and its partners had lost everything. Billions of dollars in on-paper wealth went down the tubes in a matter of weeks. So much for those carefully-tuned models.

The partners at LTCM were not cut-throat Wall Street barons. John Meriwether's inspiring leadership came largely from his kind and sincere temperament, and the other partners followed him more out of personal loyalty than greed. In fact that deep personal loyalty may even have been part of the problem, because Meriwether refused to draw hard lines with his traders. The group prided itself on living modestly, at least relative to Wall Street standards. This isn't a case of Gordon Gekkos who made money by fleecing widows and orphans. These were good, sincere people who were running an honest business. But they were led astray by the siren song of mathematical oracles, and crashed on the rocks of their own hubris.

In the past I've complained - with tongue only partly in cheek - that the problem with math is that it's too easy. I don't mean that math is easy per se. What I mean is that learning the formulas and equation is much, much easier than learning how to use them responsibly. The mathematics behind something like bond arbitrage is daunting, and if somebody has truly acquired a command of it it's easy to see how they would be left feeling pretty smug. At that point they have a very fancy hammer and the bond market is brimming with nails. They're not likely to second-guess a profitable formula. Knowledge of the math, and just the math, has only taught them enough to be dangerous.

The fact that traditional mathematical models are not very robust when it comes to the economy is not news. Economists have fretted over it, policy makers have dreaded it, and at least the good financiers have learned to take all theoretical predictions with a hefty grain of salt. There have even been popular books written about it, such as The Black Swan by Nassim Nicholas Taleb.

This breakdown comes largely from the fact that economics is prone to extreme outlier events, often called "fat tail events". The default of Russia on its debt is just such an event. They are massively impactful, and they happen rarely enough that you don't anticipate them, but just often enough that they're likely to muck up your plans. They represent a massive deviation from business as usual, usually arising from the "unknown unknowns" in a situation. I mean, who could have anticipated that the nation of

Russia would just up and default on its debts? Because of their massive debt and minimal cash, LTCM was walking on a tightrope. That's not inherently a bad thing; they were experts in tightrope walking, and had walked along this particular one many times before. But Russia's default was a massive gust of wind.

Traditional statistical models are generally based on the "normal distribution", the technical term for the classical bell-shaped curve. The normal distribution is popular partly because it's useful for describing some real-world phenomena, but mostly because it has a lot of elegant mathematical properties. If you flip a million coins, for example, the number of heads you get will be described by a normal distribution which peaks at 500,000. You are reasonably likely to get 500,005 heads, but you have no realistic chance of getting just 100,000 heads. The probability of a huge outlier like that, while technically not zero, is so small that it can be safely ignored. The normal distribution realistically guarantees that you won't have fat tail events.

Of course that's not the end of the story. The normal distribution relies on the assumption that the individual coin flips are independent of each other, which is a pretty good assumption for coins. In the real world though, those coin flips are actually home mortgages that may or may not default, or bond arbitrage deals that may or may not turn a profit. In these cases an external event, like a panicking financial market or the Russian Debt Crisis, can bias every single coin flip against you. You can still think of it as flipping coins, except that before doing the million flips you have a 1% chance of switching to a coin with heads on both sides. Now the distribution of heads will *not* be a classic bell curve. 99% of the time you will get around 500,000 heads, like usual, but 1% of the time you will have every single toss come up heads. If you're just fitting a model to historical data it is possible that nobody to date has ever gotten the biased coin, so you might not even realize this particular keg of dynamite exists. Getting the bad coin is like Russia defaulting on its debt.

Much has been made of an area of math called "robust statistics", which was developed to help deal with these problems. Robust statistics creates models that predict the existence of extreme outliers, by using alternatives to the classical bell curve. These alternatives to the normal distribution are also bell-shaped, but they taper more slowly so that the the chances of an outlier, while still small, are not negligible. Many people have jumped on robust statistics as a cure for the problem of outlier events, but I am very concerned that they're getting ahead of themselves. Robust statistics is certainly a more accurate way to describe the economy than classical statistics, but the models are not fixed. They're just less broken. *Any* description we give of the real world will necessarily be simplified to fit into the confines of the terminology we're using. Adding new vocabulary improves the situation, but it doesn't avoid the issue entirely.

The bottom line here is that adding in fancy math can't make a bad model good, any more than you can make up for ignorance in a conversation by eloquent rambling. No model contains more information than what we put into it, and that information has to be obtained the old-fashioned way. There's no substitute for plain old knowing what you're doing.

When it comes to imperfect models of the real world, my own biggest reality check came during grad school. It wasn't in anything having to do with finance, but the lesson was the same.
  While I was at Carnegie Mellon University, I did some work with modeling the performance of computer hard disks. What we wanted was a formula for how long, on average, it will take to read memory off of a hard drive. This is the research that I discussed a couple chapters ago; there I was admitting to my own fumbles in the derivation, but here I'm talking about the idealized model itself.

There are a lot of things that affect how long it takes for the computer to read a file. There's the physical shape of the disk, how fast it spins, and how fast the computer can move the reading device over the disk to where your file is physically stored. The biggest issue though is that the computer is often dealing with a backlog of other requests, for other files located at different places on the disk, and it has to figure out which order to process them in. Do you process the requests first-come-first-serve? Maybe you should process the shortest request first? One of the big time sinks is in moving the physical reading device over the disk, so maybe you should start with whichever request requires the least movement?

When trying to understand the effects of this backlog a key question is how memory requests come to the disk. They might come one at a time, with even spacing between them so that there's never much backlog. Conversely there could be long downtimes punctuated by large bursts of requests, which clog up the system like rush hour traffic. It could also be something in the middle. There is no right answer mathematically; how people use disks is an empirical question.

The conventional model for a sequence of events like this is called a Poisson process. In a Poisson process the arrival of requests is perfectly random: every instant in time there is an equal chance that a memory request will arrive, and each instant is independent of all the others. There is no tendency toward large bursts, downtimes or even spacings. Those might occur, but only by sheer dumb luck. In a Poisson process a truly massive burst of requests will be a fat tail event.

There are lots of real-world situations where poisson processes are a great model. Raindrops hitting a quarter during a storm, for example, or comets hitting the earth over time. It's not like raindrops or comets coordinate with each other on when they should strike; they are independent. However, the real reason that Poisson processes are so ubiquitous is the same reason that the normal distribution is: it makes the problems much easier (or in many cases, possible at all) to solve with pencil and paper.

So like all good ivory-tower researchers, I naively assumed that memory requests came as a Poisson process, chugged through the math and derived a formula. Now in order to publish a paper you need to at least pay lip service to the idea of real-world applications, so we got our hands on a large dataset that recorded memory requests to some real world hard drives. I wrote up a simulator to see how a disk would operate if it got those requests and compared the results to our formula.

Lo and behold, the average time lag for the real-world disk was many times longer than what we estimated. A factor of ten or so, if I remember correctly. Horrifically off! At first we went back to the drawing board with our derivation. Did we botch the logic somewhere? Was there anything obviously wrong with the final formula that might signal an error on our part? We couldn't find any problems, so eventually I went back to the data for a closer look.

It turned out that the memory requests to the disk came in massive bursts. At a given moment in time the disk was usually either idling or else working through a massive backlog - the "happy medium" of a Poisson process basically didn't exist. Every step in our derivation was sound, and the formula we derived was correct. But our simplifying assumptions about the world were spectacularly wrong. This isn't to say that the formula is useless - there are other situations in which our assumptions would hold extremely well. It's just that before using a formula like this you absolutely, positively have to check *how* useful it is, and always be on the lookout for breakdowns in your assumptions.

Note to self: always compare against real data before I try publishing a paper!

My favorite example of responsible use of mathematics in finance is Warren Buffett, one of the richest men in the world and perhaps the most successful investor of all time. It has always amused me that, even though Buffett spends a great deal of time decrying the abuse of mathematics in finance, his own empire is built upon a deeply mathematical foundation, and Buffett himself is a mathematical prodigy. His company, Berkshire Hathaway, is a "holding company", meaning that it doesn't sell any goods or services. Instead, it owns other businesses which operate more-or-less independently of the mother company. Buffett tries to stay out of the day-to-day running of Berkshire's subsidiaries; his job is to decide which businesses Berkshire Hathaway buys and sells, and at what price.

On a typical day Buffett wakes up in his modest house, heads to Berkshire headquarters and holes up in his office. The office has a window, but the blinds are drawn, and there is no computer; he doesn't like distractions. Most of his time is spent reading whatever he can get his hands on. Reports from Berkshire's subsidiary companies, newspapers, columns, and annual reports from companies he has his eye on. Buffett's job is to digest the information, synthesizing it into a big picture of the economy, in particular the parts of it he is looking at buying, and culling out the critical patterns that will dictate how they evolve.

There are a lot of different formulas for calculating how much a financial product is worth. Buffett is an expert in these various models, and the mathematics behind them. But unlike LTCM, who made money by eking small margins out of highly tuned models, Buffett focuses on rooting out the subjective, hard-to-quantify factors that break these models - but break them in a good way! If a company has stellar management, a strong brand, or a better position to weather industry changes, those factors probably won't be captured by standard accounting techniques, and the calculated value of the company can be a dramatic underestimate.

The simplest theoretical way to value a business is its "book value", which is just the sum of all its assets minus any debt. If you liquidated the company, sold everything it owns at market prices, combined that with any cash on hand, and paid off all debt, the book value is what's left over. This is a worst case scenario for what a company is worth; if the business is on a bad course and the prospects of a turnaround are bleak, the management can sell it for scrap and distribute the book value among shareholders. It rarely comes down to this for real companies, but it does happen sometimes. From an investor's perspective, if the company's market capitalization (number of shares times market price for a share) is less than the book value, then surely the shares are a bargain! It can be very hard to estimate the scrap value of all the company's assets, but if you can do it reliably, you can make a lot of money.

This used to be Warren Buffett's approach. He learned it from Benjamin Graham, the founder of this school of this so-called "value investing", and it worked well for many years. But people caught on eventually - I mean, it's not rocket science to compare a company's balance sheet to it's stock price - and there are very few opportunities like that around anymore. In fact, these days when something is selling for well below its apparent book value, there's usually a very good reason for it!

A healthy company is worth more than just the sum of its parts, so the book value is generally an underestimate of real worth. The company's assets aren't just sitting around waiting to be liquidated. They are combined into a well-oiled machine, which will be churning out profits for years to come. The last thing you would want to do is pull the plug on the business and sell everything for scrap! A more comprehensive way to value the company is to look at the future profits that it will (probably) make. How much money would you be willing to pay today in order to own those profits down the road? If you can answer that question then you have a number for how much the company is worth.

The usual way to calculate the value of future earnings is called "discounted cash flow". The key observation is that instead of investing your money in a company, whose expected profits may not actually materialize, you could instead have put your money into essentially risk-free bonds or a savings account. The only reason to buy a company's stock is if you expect it to get better returns than a savings account - why else would you accept the extra risk? As a quick note, my discussion of LTCM earlier might have made bonds sound a bit scary, but they're actually an extremely stable investment. The only reason LTCM got in trouble is that they were splitting hairs about bond prices, and then betting the house on those predictions. You can't invest much more reliably than just buying and holding US treasury bonds.

You can typically make about 5% a year on bonds. Let's say a share of stock is selling for $10 and making $1 per year. If you used $10 to buy the stock then you will make $1 per year, whereas putting it into a bond would only give you fifty cents. So if you're confident that the stock will continue to make that $1 per year, it sounds like a much better investment than the bond. The fair price of stock is $20, since you would need $20 worth of bonds to get the same profits, but you scooped up the stock at the bargain price of $10.

Of course, good luck predicting future profits! No stock will make exactly $1 per share per year forever. If you somehow knew all the future profits you could just plug them into the formula and find out what the company is worth, but that information isn't available (technically you would also need to know the bond interest rate in the future too, but that tends to be much more stable than corporate profits). If you're trying to predict profits for a young or small company you may as well be examining pigeon guts or consulting an astrologer. For a large stable company though, you can often get a pretty good estimate by just projecting current profits forward in time, adjusting them to keep in check with inflation.

When Warren Buffett finds a sound business that is selling below it's discounted cash flow value and he's confident about the future profits, he'll typically buy in. However, his favorite investments are the ones where it's theoretically selling at a reasonable price, but he has sniffed out qualitative factors that indicate the profits will grow over time. For example, unusually strong brand name, stellar management, and owning mission-critical patents all add to the likely future earnings. It's extremely hard to say how much they add, but if they are there are the company looks reasonably priced otherwise, the shares of stock are probably a steal.

Buffett doesn't only make money from seeing when models break down. He indirectly makes a lot of dough by playing the odds on fancy models, because Berkshire's core holdings are insurance companies. Insurance companies aren't selling good and services, with simple balance sheets that you can plug into a valuation formula. The whole business model of insurance is based on mathematically predicting risk, so that they can set their premiums correctly.

Yet even in a highly-mathematical area like insurance, Buffett is keenly aware of the weaknesses of mathematics, and uses that knowledge to put the odds in his favor. GEICO, Buffett's most famous insurance company, sells auto insurance. Car accidents happen a lot, so there is a lot of data to base predictions on. The losses from any single accident, though staggering for an individual, are miniscule for the company as a whole. And most importantly, car accidents are almost completely independent of each other. There's no equivalent of the Russian Debt Crisis that would make half of GEICO's customers crash their cars in the same year. This is one of those examples where a classical bell curve will work very nicely! So yes, Berkshire Hathaway is in the business of making bets based on mathematical models, but they aren't the high-rolling gambler who takes out a mortgage for a really promising hand. Berkshire is the house.

Incidentally, Long Term Capital Management once sought out investment from Buffett. In classic Buffett style the investor was folksy, encouraging, gracious, and completely unwilling to write a check. He didn't doubt the sincerity of the people at LTCM, but he was wary of the faith they put in their models. The world has a way of defying our expectations, and even though LTCM had been doing well for a few years, Buffett knew that a party founded on the minutia of highly-tuned math cannot last. A few months later the fund collapsed.

Alas, the world has not learned its lesson since the financial crash. I'm not an economist, and when it comes to the powers-that-be in the world economy I cannot comment on their understanding of economics. However, I am convinced that their understanding of math is still dangerously inadequate. I'd like to show you why I'm so convinced of that, by relating a story from my own life that still gives me the chills. It happened soon after I left grad school, when I interviewed for a job with one of the largest banks in the United States. The job was constructing mathematical models to predict something called "operational risk".

There are lots of ways a large financial institution can lose money. "Credit risk" refers to the possibility of people who owe the bank money defaulting. This includes mortgages, credit cards and the like. "Portfolio risk" comes from the possibility that the bank's assets (like stocks and bonds) could go down in price. Operational risk is a grab-bag category including any failure of the bank's internal processes. This mostly means that the bank might screw up in some way and get sued for it, but also includes such obscure events as earthquakes.

Credit risk is pretty straightforward. Like selling car insurance, every borrower has a credit history that gives you a fairly good idea of what to expect, and since each borrower owes you a finite amount of money, your losses from any one customer are capped. Furthermore, creditors operate largely independently of each other (though not completely! A bad economy can force many people to default at the same time), so it's easy to diversify by lending to many people.

Operational risk, on the other hand, is a complete mess. In the first place there is a lot less historical data to base our models on, since major lawsuits and natural disasters are a lot rarer than mortgage defaults. Operational risk is closely tied to the regulatory and legal environment, which changes from year to year. Losses are likely to come in bursts if, say, the bank implements a misguided internal policy that allows many employees to commit fraud, or a more diligent regulator enters office and discovers a lot of fraud that had already been going on. Finally, and most importantly, a single operational loss event can be extremely costly, so if your predictions are a little off the consequences can be enormous.
Operational risk includes many possible Black Swan events, like the Russian debt Crisis. Prudence dictates that a firm should always keep some amount of money on hand to deal with

operational loss, but the million dollar question (actually, much more than a million dollars) is how much.

Considering the whole world-wide financial meltdown thing, regulators thought it imprudent to leave that decision to the banks and trust in their good judgement. They needed an objective, enforceable criteria for calculating the reserve level. Intuitively they wanted it to be enough money to cover a realistic worst-case scenario. No alien invasions or zombie apocalypses, but other than that banks should have enough money on hand to weather any storm.

The criteria they chose, at least for the largest and most complicated banks, was as follows. The bank would construct a mathematical model of how likely they were to lose how much money from operational breakdowns. This intricate model would be trained on historical data, and would tell you how likely any given range of losses (like, between 1 and 2 billion dollars) was. Once you have that model in-hand, the probability of losses exceeding reserves must be less than 0.1%. This reserve level is often called the "value at risk", or the VaR.

Once you have a statistical model on hand it's a straightforward problem to calculate the VaR. But once you've calculated it, what does that number actually tell us about the world? Unfortunately not a lot. The clean precision of a statistical model masks the extremely messy process of fitting that model to the real world, especially for something as complicated as operational risk. The data you're fitting to will be woefully incomplete, and the model will probably be brimming with fudge factors. We can look at losses for the last few years to get an idea of what a "typical year" looks like, but if we look back further than that the economy and regulatory environment will be completely different. So there is no historical precedent for telling us what a "really bad year" would look like, let alone a one-in-a-thousand bad year. You can certainly pay somebody to fit a model to the data and then calculate the VaR, but it's garbage in garbage out. The data we have on hand is good enough to estimate a typical year, but it's useless for estimating extreme outliers. Even reliably estimating the one-in-a-hundred bad year is nearly impossible. The idea of calculating the one-in-a-thousand bad year is just silly.

I went into the interview figuring that regulators had chosen the 0.1% cutoff so that politicians could sound "tough on the banks". I expected that people at the firm would work hard to construct the best models they could, but it would be something of an inside joke that regulators and politicians thought the numbers were reliable. So I was dismayed to find that the bank's analysts thought that the VaR could be well approximated. They knew the problem was *hard*, so they were throwing a lot of advanced math and fancy degrees at it. They failed to realize that the problem was actually *futile*, and the numbers they were deriving weren't an approximation to anything. There's no way to work around the problem of not having enough data.

In a model that's trained to a "business as usual" market, the outliers will represent statistical fluctuations in the normal loss levels. Garden variety loss events, and by dumb luck there were more of them than usual, or the ones that happened were unusually bad. But I can guarantee you that the 0.1% bad year will be defined by something very different from business as usual. Maybe a natural disaster on a massive scale, an Enron-style accounting fiasco, or a computer bug that enables thousands of electronic frauds. Who knows what it would be? I certainly don't. Neither did the analysts at the bank, but somehow they still thought they could quantify it.

To be clear here, I'm not criticizing the regulations. If you're going to police the industry then you need some kind of objective standard, and VaR is as good as anything I could come up with. However, when it comes to actually predicting one-in-a-thousand-years disasters, you may as well be examining pigeon entrails.

The analysts at the bank were constructing reasonable, defensible models of how the bank loses money, using the best available tools from statistics and computer science. They were doing all of the math right! What they didn't understand was the tenuous connection between their models and the real world. The numbers they produced weren't just for regulatory compliance; people wanted to use them for making business decisions. The fact that something is a number has some psychological overtones of precision and confidence. Many people who are calling the shots in the modern economy, even after being stung by the 2007 economic recession, cannot see past the number to the severe limitations of the whole endeavor. You can't calculate the strength of a poker hand if you don't know what cards were in the deck in the first place.

I declined the job.

9 THE NEW AGE OF EMPIRICAL MATH

A theoretical physicist and her friend go to the horse races one day. Neither of them knows much about horses, so to make things a little more interesting they decide to compete with each other. They each start off with a hundred dollars for betting on the races, and decide that whoever ends up doing the worst has to buy dinner in the evening. The friend takes a look at the starting lineup, shrugs her shoulders and places her first bet. The physicist smirks to herself, pulls a pencil and paper from her purse, and starts breaking out the equations.

At the end of the day they reconvene to compare results. The friend has won some races and lost others, eventually ending up with a hundred and ten dollars. The physicist though has been completely wiped out, with nothing to show but discarded paper.

"My gosh!" exclaims the friend. "How did this happen? I don't know a thing about horses, so I just bet on whichever one looked the spunkiest. But you put all that work into it, and this still happened. What on Earth were you doing in those derivations?"

"Well", explains the physicist sheepishly, "I started by assuming a spherical horse..."

This is one of the classic jokes among physicists. Seriously - some variant of the spherical-horse joke is a rite of passage in every self-respecting physics department. It's our way of poking fun at the drastic simplifications you often have to make in order to solve a problem with paper and pencil. Otherwise the math is just intractable. As the joke shows though, sometimes these simplifications lead us into absurdity.

Ok, so what's a better way for our physicist to go about solving the problem? Deriving a formula for how fast a horse can go requires too many simplifications, and we'd like to do something a little more rigorous than eyeballing which horse looks the "spunkiest". For most of human history though those have been the only games in town. Either you drastically oversimplify the situation so you can solve it with paper and pencil, or you make a guess based on (hopefully) expert knowledge.

Starting in the 1700s or so there was another option, useful to some degree in some cases. This was the field of statistics. It could give you the objectivity of a derivation combined with some actual knowledge about horses. For example, you could test rigorously to see whether a horse with longer legs runs faster, or what type of food makes them more athletic. You could compare the track records of various horse judges, and see whose intuitions actually played out on the race track. But crunching the numbers was a long, laborious process for even a single one of these factors.

Those days are now over, and good riddance to them. There is a new kid on the block, a radically new approach that has the best of all worlds: the objectivity of derivations, the breadth and detail of a domain expert, and the ability to actually get an answer in hand. What is it you may ask? You use a computer.

In the case of horse races you would probably use something called "machine learning". You feed the computer all the data you can get your hands on about previous races, and the specs on the horses running in them. What the horses ate in the morning? Throw it in. The temperature that day? Absolutely! The number of siblings the jockey has? Well sure, I guess you can add that in too. What's the harm?

The computer will sift through all of these tens, hundreds, even thousands of different variables, checking each one to see whether it correlates with the race outcomes. It can even look at combinations of the variables. Maybe the temperature is a significant factor for younger horses, but not for older ones? Often no single factor can tell you much, but in aggregate the little scraps of data form a powerful tool that can tilt the odds overwhelmingly in your favor. If our physicist had realized the futility of her pencil-and-paper derivations and just trained a computer to place the bets for her, she would probably have come out ahead of her friend.

Machine learning algorithms like this are what filter spam out of your email and pick ads for you to see when you visit a website. They trade stocks and bonds, forming the life blood of financial markets. They even recommend movies to you on Netflix. In the future they will drive your car and monitor your health, and who knows what else. All of this can be done seamlessly, in fractions of a second, for people all over the world.

There are many different types of machine learning algorithms on the market. What they all have in common though is that they use computers to do things you never could by hand. Adding up a million variables or checking a thousand logical steps perfectly are impossible for a human, but child's play for a machine. This brute power lets machine learning algorithms search for patterns empirically, through what amounts to a gargantuan game of guess-and-check, rather than being constrained to what's possible with pencil and paper.

Machine learning isn't the only option our physicist had. Instead of assuming horses were spherical and trying to derive a formula for how fast they can go - presumably as a function of the horse's weight and diameter - she could have put together a 3D model of exactly how each horse in the race was shaped, and then used a computer to simulate how air would flow over it while galloping. Trying to derive a formula for how air flows over a complicated shape is generally an exercise in futility, but simulating airflow with a computer program is a synch. In fact that's how engineers design aerodynamically sound planes. In place of a general-purpose formula for all animals this approach gives an approximate speed for each individual animal. And that's really what you wanted in the first place, isn't it? For this problem a formula is just a means to an end; what you really want is a number. Why pull your hair out looking for an exact formula - which would only be exact if the horse were spherical - when a few minutes of supercomputer time will give you what you need?

What is Math?

That's really the heart of the matter. Formulas, derivations and rigorous proofs are fantastic if you can get your hands on them, but there are many situations in which they are impractical or even impossible to make work. Maybe Albert Einstein would be able to derive a general formula (more likely though, even he couldn't do it), but all you actually need in this case is a good approximation.
The reason that by-hand derivations have become so dominant in the world of math isn't that they're more useful, correct or elegant. It's just that you can do them with pencil and paper, and until about sixty years ago that was the only way math could be done.
The physicist in our joke truly needed to assume a spherical horse if she wanted to make her derivations tractable. But tractability is not required if you're happy with an approximation that's good to five decimal places.

Now, using a computer to get approximate numbers is its own can of worms, of course. For airflow simulations you still need to figure out what shape each horse is, the best way to plug that shape into a computer, and how to run the simulations themselves.
Machine learning requires that you hunt down different data sources, carefully vet the accuracy of the trained computer, and make sure that the patterns it finds aren't just flukes. There is still a lot of legwork to do. It's just not in pencil-and-paper math.

In a situation like this the computer isn't proving theorems, doing derivations, or giving us any theoretical truths about the world.
The discoveries it makes are empirical facts. There are theorems, but they mostly describe properties of the algorithms used. For example, one simulation method might complete in fewer steps than another one. Or maybe one machine learning algorithm handles outliers in the data in a more robust fashion. Solving these problems is an art and a science, a dance between abstract theorems and empirical realities.

This is the future of mathematics, where human intelligence dovetails with the brute empirical power of computers to solve problems that neither could tackle on its own. I believe that in the coming decades real-world math will become a blend of theorems with rules of thumb, equations with algorithms, and human insight with raw computation. In fact this change has already happened, and it has opened up new vistas of fascinating, useful discoveries.

In this last chapter, I will try to give you a small glimpse of the awesome things that are afoot. There is no possible way I could really do it justice, but I hope to at least whet your appetite. If at the end you are not grinning with excitement, then it will be my fault for not conveying how awesome this stuff is.

This isn't really a new direction for mathematics though, but a return to its roots. For the last two thousand years math has been dominated by the theorems and derivations that are the legacy of the Greeks. This austere, formal type of math has served us well. It's what let Newton and others create modern science. It enabled the industrial revolution, and has played important roles in our philosophy and art. But before it came a different kind of math, one that was more empirical, down to earth, and organic.

For Hera, our first paleolithic mathematician, numbers were just a way to keep track of the world. She certainly didn't know what a proof was, and never saw a derivation in her life. At most she understood basic arithmetic. "Math" was probably not even a thing in itself for Hera; it was just some jargon for keeping track of the world around her.

Thirty thousand years later, our Egyptian mathematician Ram had a more sophisticated version of the same idea. Math had evolved into a specialized skillset, rather than just a common jargon that everybody used, but other than that it was as real-world as ever. Ram crunched numbers to brew beer, erect buildings, navigate ships, balance books and chart the stars. He and his ilk enabled the rise of complex modern civilization by focusing on practical problems and doing whatever was needed to solve them. Their methods were sometimes crude, and sometimes only approximate, but they got the job done.

Ancient Greece is where there was a course correction, and mathematics morphed from an unwieldy offshoot of human language into a sleek extension of human reason. The philosophers wanted not just results, but understanding. True knowledge rather than rules of thumb. Under their tutelage, abstract mathematics acquired a life of its own, fueled by logical deduction instead of practical needs or real-world discoveries. Of course people didn't just stop caring about applications entirely, but "pure math" became a parallel discipline that didn't depend on the real world to justify itself.

In the nineteenth and twentieth centuries rigor was taken to its grotesque, logical extreme, separated first from physical reality, and ultimately from the very notion of meaning. Researchers reached the bottom of the rabbit hole, and found that it was a rat's nest of eerie paradoxes. The most important discovery was the fundamental inadequacy of formal math. It could never give us all truth, and instead was plagued by the same limitations as any other language.

Now, at the cusp of the twenty-first century, we have come full circle back to empirical math. Armed with the power of computers we are venturing into territory where rigor alone cannot take us. It is a fascinating wilderness full of truths that can't be proven, equations that can't be solved, and applications that you have to see to believe.

You've probably seen the following plot in a movie or TV show. Somebody travels back in time, after being given a stern warning not to change anything. When he gets to the past though he is overwhelmed by the wonder of it all and, while walking around gawking, accidentally steps on a leaf. At first he's a little nervous since he wasn't supposed to change anything, but how big a deal can one little leaf be? He then returns to the future only to find that the whole world has changed. Maybe his family has been ruined, he himself had never been born or, depending on how far back in time he travelled, the human race never even evolved. Oops...

This is called the "butterfly effect", which is sometimes summarized by saying that a butterfly flapping its wings in America can make it snow in China. The idea is that in a highly complex system - like, you know, the world - a tiny difference in how it starts out can balloon into massive, unpredictable changes down the road.

The idea of causing a snowstorm in China is charmingly apt, because the butterfly effect was discovered in the course of trying to predict weather patterns. The weather is of course notoriously hard to predict, and in the 1960s a meteorologist named Edward Lorentz was trying his hand at it. Lorentz took the standard physicist's approach to the problem: oversimplify it to the point of absurdity, then cross your fingers and try to solve the equations. Cutting out everything that even smelled extraneous, he condensed weather systems down to just three parameters, along with three equations that describe how those parameters evolve over time. Of course he comically oversimplified how the actual atmosphere works, but he hoped to at least be able to pull out some salient patterns.

The specific part of weather that Lorentz was interested in was convection, where you have currents of hot air rising (and cooling off as it rises) and cold air descending (and warming up as it descends). The three parameters that Lorentz used to describe convection were called x, y and z. Roughly speaking, x measures the strength of the convection currents and y measures the average temperature difference between the air going up and the air coming down. The z parameter is a little bit more complicated: it measures how smoothly the temperature transitions between the cool air higher up and the warm air lower down.

Without further ado, after all the simplifications he factored in, here is Lorenz's system of equations:
```
rate of change of x = 10 * (y-x)
rate of change of y = x * (28 - z) - y
rate of change of z = x*y - (8/3)*z
```
The numbers 10, 28 and 8/3 that I've plugged in actually have different values for different weather systems, but they are fixed

for any single system. I picked these particular values since they give rise to the phenomena I'm going to be describing.

Don't worry about what these equations mean physically. If you stare at them long enough they can start to seem plausible, but mostly people don't waste their time with it. I'm actually being unconventional by even mentioning that x (sorta-kinda) measures the strength of the convection currents! Move away from the real world for a minute and just think about the system as three numbers that are changing over time. If you happen to know what those numbers are at any point in time, then you can plug them into these equations to see how things are evolving at that moment. Knowing how fast things are changing tells you what things will be like at the next moment, and so on.

Lorentz's initial hope was to derive a solution to these equations. He wanted a tidy little formula that would give you the values of x, y and z at any future point in time, if only you knew how they had started out. He labored over the deceptively simple equations with pen and paper, but couldn't make any headway. Finally, rather than giving up, he decided to try something different. He plugged the equations into a computer and used it to estimate, instant by instant, how a weather system would evolve over time. Any simulation like that will only be an approximation, since the tiny errors start to pile up as the simulation steps thousands of moments into the future. But hey, Lorentz must have thought, these equations are already just approximations to actual weather anyway, so what's the harm in seeing how it goes?

Well, it turns out there's quite a bit of harm. Here is the monstrosity that came out of Lorentz's computer:

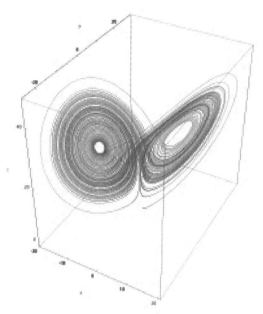

[credit: http://fr.wikipedia.org/wiki/Fichier:Lorenz_attractor_boxed.svg]

This graph is what happens if you treat x, y and z as coordinates in 3D space and plot out the way the system evolves. Of course the parameters aren't actually physical coordinates (remember, they're things like the temperature difference between different air currents), but pretending they are is a useful way to visualize what's going on.

In this case, the system's overall path is shaped roughly like a deformed butterfly. There are two "wings", where the path spirals around in an orderly fashion. Physically, spiraling around on a wing means the weather system is varying in a more-or-less predictable cyclic pattern. There are two such patterns that are possible. The wings intersect each other in a hopelessly tangled rat's nest that makes up the butterfly's body. If the system starts out on one of the wings it will spiral around that wing gracefully for a while, meaning a predictable weather cycle. Eventually though it will float off into the butterfly's body, where it will jitter around spastically before returning to one of the wings.

Which wing it goes to is almost impossible to predict. It could go back from whence it came, replaying the previous cyclic weather, or dart off to the other wing in an about-face for the real-world weather. In the chaos of the butterfly's body, something as small as the flight of an insect can change which wing it ends on.

There is no tidy little formula for how this system evolves over time. There is no hope, prayer or approximation for such a formula, or a long and complicated formula for that matter, no matter how much a pencil-wielding mathematician might want one. We're not in Euclidean Kansas anymore! If you want to know how this system changes over time then your only option is to simulate it with a computer, like Lorentz did. As your simulation progresses tiny errors - from the way that computers round off numbers, from the fact that you can't use infinitesimally short instants, and so on - will accumulate and throw off your predictions. Eventually you'll have no idea whether your simulation is even on the right wing. *This* is the butterfly effect!

The bottom line about accurate, long-term weather predictions is this: abandon all hope. However, you still might be able to say something useful. By running thousands of simulations you can see how long the system tends to stay on one wing, i.e. how long we can expect the weather to stay on a predictable cycle. You can also ask when in a cycle the change is likely to take place, and other empirical questions like that. For practical purposes, your simulations start looking less like an abstract math problem and more like scientific data that was gathered out in the field.

There are also some things that you can prove about the system. Even though you can't derive a formula for what the parameters will be in the future there are still some guarantees that can be made. For example, you can prove that the path stays confined to a certain region of 3D space, rather than shooting off to infinity or anything like that. Good luck figuring out where it is within that region, but there are certain absolute boundaries that it must respect. This theorem shouldn't be surprising, since real weather systems don't spontaneously blow up into continent-wide hurricanes. It's nice to see that this feature of real weather systems has made it into our model, and it's nice to be able to prove that rigorously.

The deformed butterfly that Lorentz found is now called a "strange attractor", and it turns out there's nothing special about it.
Lorentz's model isn't some pathological edge case. Many real-world systems exhibit this same kind of unpredictability, and it turns out to be a rare luxury when you can actually solve the equations. Models in biology, chemistry, and even economics have their own strange attractors, which defy all formulas and force us to study them empirically. Chaos, it turns out that, is truly the nature of the world.

A very hot area right now, one whose complexity dwarfs anything Lorentz could have imagined, is biology. Even a single cell is a monstrously complicated entity, with digital and analog components and a wealth of intricate moving pieces, and we're just starting to figure out how it works. Lorentz found chaos lurking in three innocuous little equations, but in biology even the models themselves are monstrously complicated.

"Systems biology" is an emerging discipline that tries to understand biology not by examining its building blocks, but by seeing the processes as a unified whole. For example, instead of just looking at the structure of the insulin molecule in excruciating detail, systems biologists might describe the feedback loops that insulin has with blood glucose and other regulatory hormones.
The resulting model would be similar in spirit to Lorentz's model of the weather: a collection of parameters that evolve over time. In this case the parameters would mean things like the concentration of blood glucose, rather than air temperatures, but as abstract mathematics it's the same idea. Could the glucose-insulin system have its own "wings", like the Lorentz attractor? Might one of these wings correspond to healthy functioning, while the other is a pathology like diabetes?

For the systems biologist, an organism is a massive network of feedback loops. They dovetail with each other, with the same molecule being an actor in multiple different processes.
Sometimes these feedback loops step on each others toes. Many illnesses, like breaking a bone, come from malfunctioning body parts or external pathogens. But other ailments, we are now learning, are system-level pathologies. The components are all fine, but the system as a whole has gotten into a destructive loop. This more subtle category seems to include diabetes, alzheimer's, autism, and Crohn's disease, to name only a few.

Scientists are now starting to study biological processes in the same way that Lorentz studied the weather. Something like the immune system, the rhythmic pumping of the heart, the digestion of food or the metabolism of a drug all fall under the same umbrella: a set of parameters and some equations that say how they change over time. These models are all idealizations of actual biology, but even as simplified as they are they still resist pencil-and-paper solutions. Systems where you can actually derive a formula are the exception, not the rule.

A critical way that systems biology differs from Lorentz's systems, besides just having many more parameters, is the presence of DNA. The weather is a continuous process because temperature, humidity and air current vary smoothly from one place to another and one moment to the next. You don't take one step and go from a calm sunny day into the middle of an arctic hurricane! DNA though is fundamentally digital, in the same way that computer data is built up out of bits that can be either 0 or 1. In order to really understand biology we'll have to build hybrid models that combine continuous parameters, like blood glucose levels, with digital parameters like our genetic code. For the time being though, there's a lot of low-hanging fruit just from studying DNA on its own.

In the year 2000 the Human Genome Project published a rough draft of the entire human genetic code. This was a stupendous achievement, but in itself it was actually pretty useless. The genome isn't composed of diagrams and explanations, which are understandable and can be turned directly into medicines and bioengineering. It's more like getting your hands on the 0s and 1s inside of a computer, but not having any idea how they're organized into files, whether those files are pictures, videos, word documents or something else, or whether one file might be broken into several pieces. And of course, not knowing what else might be encoded besides files. The genome may be the "blueprint of life", but it's a blueprint designed for a very non-human architect, whose mind we don't understand and who seems to have a love of complexity and edge cases.

DNA is made of molecules called nucleotides, strung together into long, twisted ladders. A single ladder can be many millions of nucleotides long - humans have 42 such ladders, which usually called "chromosomes". The nucleotide molecules come in four types: adenine, tyrosine, guanine and cytosine, which are usually abbreviated with the letters A, T, G and C. The order of these nucleotides on a DNA ladder constitutes the genetic code. There are 3.3 billion of these letters in the whole genome - roughly the length of a thousand copies of War and Peace. But unlike Tolstoy's epic work, they aren't composed into words, sentences and chapters.

Before the genome was sequenced scientists thought they understood the big picture and were just filling in details. Their so-called "Grand Dogma of Biology" says the following. A chain of DNA is broken into stretches called "genes", which you can think of as the equivalent of files in a computer. Each gene encodes the structure of a protein molecule, in the same way that you might use computer bits to encode a video. In order for the cell to manufacture the protein, the gene first gets translated from DNA into a very similar molecule called RNA. The RNA carries the same information as the original gene, but it is free to float around in the cell. When the strand of RNA eventually floats over to where proteins are assembled, it gets used as a template for constructing a protein. The various proteins then do everything else that goes on in the organism, including all the hoopla about managing the DNA and RNA. That's the big pictures: DNA gets translated to RNA, RNA gets translated into proteins, and then proteins do basically everything else. When biologists first began exploring the genome they expected to find that it consisted of a bunch of genes for proteins, together with some "syntax markers" that showed where one gene ended and another began.

This picture isn't wrong, but it's woefully incomplete. Indeed there are many genes in the genome that code for proteins, and syntax markers that say where they begin and end. However, it turns out that only about 1.5% of our DNA actually codes for proteins. What the rest of it does is a mystery. In some cases the DNA gets translated into RNA, but the RNA molecule does something useful in itself, rather than being a template for proteins. Certain stretches of DNA are never translated at all, and seem to be there to stabilize the DNA molecule itself, causing it to twist and bend in convenient ways. Other stretches of the genome seem to have been used by our distant ancestors, but are no longer actively used by us. There are dormant genes in chickens, for example, that when artificially activated give them teeth like their dinosaur ancestors.

The ways that the genome can vary between individuals are staggering. Let's go back to the picture of a thousand copies of War and Peace. The variations that scientists expected, places where the text differs by changing one letter, are very common. There are around 50 million of them. Yet it turns out that they are only a small slice of the variation. There are other places where whole words and sentences are missing from some copies of the book. Or the order of two sentences can be switched. In other cases you can have multiple copies of the same sentence, with subtle variations between them. Autism, for example, seems to have less to do with what genes are in your genome and more to do with how many copies of them you have. What happens when you start re-ordering whole chapters? Or referencing particular parts of the book within the text itself?

Many people I've worked with devote their professional lives to looking for ways to identify these patterns. Working with a fundamentally digital system like DNA is very different from simulating continuous systems like weather or blood glucose.

However, it's equally reliant on computers, and the same philosophy of studying systems empirically carries over.

Biologists don't generally prove theorems or write out derivations, but they do make monumental discoveries.

One pattern in the genome that I personally am fascinated by is the effects of viruses. Many viruses attack a cell by inserting whole new chapters into the host genome, which reprogram the host cell to act as a factory for more viruses. If something goes wrong and the chapter doesn't get inserted correctly the cell might survive, and the new DNA amounts to a massive mutation. If the new mutation codes for uncontrolled cell growth, rather than uncontrolled virus production, the cell becomes cancerous.

Cervical cancer is caused almost exclusively in this way, by the human papillomavirus (HPV). I predict that when we have sequenced the genomes of more cancerous cells, and developed better algorithms for making sense of them, many more cancers will turn out to be caused by virus attacks gone wrong.

Most of the time these things won't be cancerous, and if a virus mutates a cell in an ovary or a testicle it might get passed on to the next generation. This is one of the most powerful mutation mechanisms for driving evolution. Inserting viral DNA is not a tiny random tweak, or a selective pressure on a population. It's introducing, all at once, dozens of new genes that already code for meaningful proteins. This, I think, is where the really big leaps in evolution often come from. Those extra genes could introduce totally new functions into an organism in a single generation.

More likely, the new DNA will lay dormant for generations, at some point mutating a bit and becoming useful. This may sound like an alarming hypothesis, but almost 8% of the human genome has already been traced to a virus family called HERVs. It was placed there randomly by viral attacks gone wrong, deep in our evolutionary past. Surely we carry other viral DNA as well. It's just mutated so much over the eons that we can no longer tell where it came from.

Modern economics has been described as an employment program for good mathematicians. As a physics guy myself, a huge number of my friends and classmates ended up working on wall street, and essentially all of us have considered jumping into finance at some point.

In the old days stocks and bonds were sold in person, and to be a securities trader was essentially to be salesman. Nowadays though, almost all trades are done by machines. A modern trader is more likely to be found sitting in front of a computer, downing coffee and staring at charts that tell him how well his algorithm is performing in real time.

These algorithms hybridize theoretical rigor with empirical facts. In many cases they are based on machine learning, trained on historical data to recognize profitable patterns. The price of stock A is going up? Well, that's often followed within 20 seconds by a rise in stock B, so we should buy some now, hold it for a minute, and then sell. In other cases the algorithms are based on mathematical certainty: they are intended to make a profit, but guaranteed to not have too big a loss. Some algorithms have both: they are optimized for current market conditions, but have fail-safes built-in in case those conditions change.

When automatic trading first became popular there were colossal sums of money to be made. Financial markets were filled with small, systematic inefficiencies that could be exploited. Some of them had simply never been noticed, and others were so transient that they disappeared before anybody could trade on them. Computers, which could comb through gigabytes of historical data and make decisions in fractions of a second, solved both of these problems. Scientists and engineers flocked to finance in a massive gold rush.

They're still there today, and still getting hefty pay checks, but times aren't as sweet as they once were. The inefficiencies they used to capitalize on have largely vanished in the major markets, as more and more traders have caught on. Third world countries provide fresh meat, but those opportunities quickly disappear as well. The inescapable fact is that more and more firms are duking it out over a roughly fixed-size pie.

My prediction is the following. There are only a finite number of basic strategies that you can trust a computer to execute, and a finite number of areas they can be applied to. At some point all these strategies will have become general knowledge. As this happens, the competition will shift toward buying the best computers, and making sure that your algorithm runs the fastest.
We are already seeing this arms race happen, as firms vie with each other to shave milliseconds off their algorithms, in the hopes that their computer will be at the front of the line when a deal shows up. There are limits to this too though. Within a couple decades purely algorithmic trading will have condensed into a handful of highly technology-focused firms, who eek out tiny margins across millions of small transactions. They will make a profit, but they will function more like utilities companies than the money factories of old. The glory days of algorithmic trading will be over.

For most companies, what will replace purely algorithmic trading is a situation where people and machines work hand-in-hand. A human will use computers and their own knowledge to look for possible opportunities in the market. They will then evaluate those opportunities using both their own analysis and sophisticated computations. Eventually the human will tell a computer to "buy ten thousand shares of IBM today". The choice to buy or sell is a human's judgement call, but once the decision has been made the computer will execute it on its own. The computer will break the order up into many smaller transactions, spacing them out over the day. It will shop around for the best deals on each trade, possibly using machine learning to sniff them out. This automated bargain shopping will get a slightly (say 1-2%) better deal on those ten thousand shares. It's not enough to build a business on, and certainly not enough to make up for a misguided purchase, but it's an excellent bonus if you can achieve it consistently.

I can't talk about the future of math without talking about physics. Hopefully this book has convinced you that physics is not the end-all-be-all of mathematics, but the fact remains that physics problems have driven a huge amount of mathematical development from the scientific revolution up to the invention of computers. Nowadays computer science, biology, economics, and a ton of other areas are generating their own new math, but physics hasn't scaled back.

The two great achievements of twentieth century physics are quantum mechanics and general relativity. They are both reductionist theories, in that they try to explain as much of the world as possible in terms of a few underlying, fundamental laws. With pencil and paper they let you derive the shape of a hydrogen atom, or the critical mass for a star to collapse into a black hole.

However, as physicists have started to apply these theories to more complicated situations they have run into a snag. Let's say you work at Intel, designing high-performance transistors. The laws of quantum mechanics might tell you, in principle, everything there is to know about how a semiconducting material works. Kudos to you, but that's not very useful if you can't solve the equations! The same thing we saw with Lorentz applies here: when you're tackling real-world problems, equations that can actually be solved with pencil and paper are the exception, not the rule. So if you're that engineer working at Intel, you plug the laws of quantum mechanics into a computer, along with the specs for the semiconductor you're studying, and out pops a numerical approximation.

If you want to understand the formation of planets or the shape of a complicated molecule, you once again have to simulate it on a computer. If you derived an equation that can't be solved, you'll have to settle for an approximate solution.

When I go to physics seminars nowadays, a growing number of them are given by people who, realistically, spend their time writing computer code, and probably haven't done a proper derivation in years. Sometimes these researchers are made fun of for being code monkeys in scientists' clothing, since "real" physicists do derivations. But those criticisms are unfair, because these people don't actually care about computers. In the same way, Einstein didn't really care about equations; it's all just a means to an end. What real physicists care about is physics, and they'll use whatever tool is expedient to make sense of the world.

Reductionist physics is still going on, where people are struggling to combine quantum mechanics and general relativity into one overarching theory. However, the real meat of the discipline has moved away from reductionism. It's no longer about unifying disparate theories, but about seeing how those theories play out on a grand scale.

Given all of this, it's astounding that most students don't know what the word "algorithm" even means. Computers are treated as tools for checking email or researching a new topic on the internet, but they aren't seen as a paradigm for viewing the world. The ability to work with basic algorithms, a skill which is bleeding into every discipline under the sun, is treated as a special topic for advanced students nearing the end of high school (personally, I wasn't really exposed to it until my senior year), rather than a foundational skill for all knowledge workers. I'm not an educator, and I don't want to pretend that I know how best to teach students. However, I do want to at least give you some flavor for what I'm talking about, and show you how algorithmic thinking can actually be a very natural skillset.

Let's go back to the Lorentz system for modeling weather patterns and see how you could write an algorithm to simulate it. Recall that Lorentz simplified atmospheric convection down to just three parameters x, y and z, governed by the following equations:
```
rate of change of x = 10 * (y-x)
rate of change of y = x * (28 - z) - y
rate of change of z = x*y - (8/3)*z
```
If you know what x, y and z are at any point in time, then you can plug them into these equations to see how fast the system is changing.

For example, let's say that right now x, y and z are 1, 2 and 20 respectively. Then:
```
    rate of change of x = 10 * (y-x) = 10 * (2-1) = 10
    rate of change of y = x * (28-z) - y
                        = 2 * (28 - 20) - 1 = 15
    rate of change of z = x*y - (8/3)*z
                        = 1*1 - (8/3)*20 = -52 1/3
```
So in a minute x and y will be a little bit higher than they are right now, and z will be a little bit lower. Since we know how fast the parameters are changing we can estimate what their new values will be after that minute. Plug these estimates back into the formulas, and we have an estimate for how fast the system *will* be changing a minute from now.

Conceptually you can go on like this for as long as you want. If you do it 60 times, you'll have an estimate of what the weather system will in an hour. Every timestep you simulate introduces a little bit of error, since the parameters aren't changing at a constant rate over the minute, but if you're careful you can extrapolate a long ways into the future (people spend a lot of effort in figuring out clever ways to be careful, and in figuring out just how accurate a simulation is).

So now we've figured out what our algorithm is. Here's what it looks like when we break it down into numbered steps:
```
Input: starting values for x, y and z
       size of time step   (like, 1 minute)
       number of time steps to simulate
Algorithm:
    1. x, y, z = their starting values
    2. num_steps_done = 0
    3. rate_x = 10 * (y-x)
       rate_y = x * (28 - z) - y
       rate_z = x*y - (8/3)*z
    4. x = x + rate_x * time_step
       y = y + rate_y * time_step
       z = z + rate_z * time_step
    5. num_steps_done = num_steps_done + 1
    6. if num_steps_done = number of steps to
simulate: output x, y and z
    7. otherwise, go to step 3
```

If this seems pretty simple to you, then you're right. It is! The Lorentz equations themselves are messy, and they're hell to derive in the first place, but once you have them in-hand it's straightforward to estimate how they change over time.

This process of breaking a problem down into an algorithm that solves it is the essence of what I call "algorithmic thinking". In the example with Lorentz you understand what you want to do and roughly how to go about it. If you sat down with pencil and paper you could work out the first few steps by hand. But it becomes a different game if you want to solve the problem with a computer. Computers don't "understand" the problem in the way that you do; they just do what you tell them. So you have to break the process down into rote steps that the computer can follow exactly.

Speaking from experience, it can be surprisingly hard to break a process you understand down into a precise algorithm. The human brain tends to think about things in pictures, heuristics or abstract relationships. Going from these to concrete steps can be a major mental feat, and algorithmic thinking takes a lot of practice to really master. From what I've seen it's usually the biggest stumbling block for humanities people trying to get into programming computers.

Ok, you might be saying. I get it: computer programming is useful for simulating the Lorentz equations, so algorithmic thinking should be taught to those hordes of students who want to study the details of atmospheric convection... Actually this skillset can be equally useful outside of science classes. For example, you can have digital design students create simple games like minesweeper, which have logic-based algorithms at their heart. English students can analyze Shakespeare's plays to see how many of the words he uses no longer show up in the dictionary. They can also compare the average sentence lengths of various authors, or the number of unusual words they use; does Hemingway *really* write in a simpler way than Jane Austen? Even in P.E. students can write algorithms to analyze baseball statistics or calculate how many calories they've burned. Computation should be treated as a general-purpose tool, rather than a nerds-only special subject.

Again I am not an educator, but I do know that computers are fast becoming an indispensable skill for knowledge workers. The ability to break down a problem to the point where a computer can solve it is at least as important as the ability to do math beyond arithmetic, and it deserves a corresponding place in the curriculum.

If you followed my discussion about simulating weather systems in the previous section then congratulations: you understand multivariable calculus! I dressed it up a little differently from what you would see in a formal calculus class, but the content is exactly the same. In technical jargon, the rate at which the x parameter is changing is called the "derivative of x(t) with respect to time", the equations for those rates of change are collectively called a "nonlinear dynamical system", and the algorithm we wrote down is a type of "numerical integration". It's intimidating jargon that masks simple ideas. In the normal curriculum mathematical students get exposed to this material about midway through their college career, and other students never see it at all. Then why, and more importantly how, am I able to include multivariable calculus in a book for a general audience?

Calculus is an example of math that is easy to understand, but very tricky to implement. Using a computer to do calculus approximately can be very simple, as we saw. But doing calculus exactly, with pencil and paper, becomes quite involved. Applying calculus operations to simple formulas often yields complicated, messy results. Students working with fractions are likely to run headlong into trigonometry. Statistics ends up blending into geometry. And of course, everybody is up to their ears in difficult algebra.

Because calculus derivations so quickly become complicated and difficult, the educational community has decided to put off teaching calculus until the end of most students' careers. At that point it can be applied to everything they've seen (and/or forgotten) in the last twelve years. But the underlying concepts of calculus can be understood by kids in elementary school - I know this personally, having taught it to a number of them myself!

In my mind calculus classes should be re-tooled to incorporate more computer algorithms, introduced earlier in the curriculum and integrated with other subjects, but that's just the tip of the iceberg. Mathematics in general is suffering from the legacies of formalist math, the days before computers were invented, and a time when we weren't a knowledge worker society. Throughout the curriculum we need to re-align subject matter with how the human brain actually thinks about problems, and with the reality of what skills are and aren't useful in the modern world.

The most controversial subject that's potentially on the chopping block is doing arithmetic by hand. The great irony here is that arithmetic actually comes very naturally to the brain. Children easily understand what addition and subtraction are, and they will even invent their own methods for doing it efficiently on their fingers and toes. Multiplication and division are a little more involved, but still very intuitively accessible. These concepts are not hard. Students only start to struggle when we begin teaching them rote algorithms - carrying the 1, long division, etc - for solving the problems. This kind of meaningless symbol manipulation simply doesn't come naturally to the brain. Rather than indulging children's ability to move on to more exciting topics (like geometry and calculus) we hammer them with rote exercises until they've mastered the critical skills that they... oh wait, nobody does math by hand anymore.

I am both a working mathematician and a (more or less) normal human being, and in neither capacity do I ever actually do arithmetic by hand. What I do all the time though is approximate calculations, and I think that's the direction arithmetic education needs to go. What's the more useful ability: exact long division by hand with pencil and paper, or approximate arithmetic in your head? If you're planning a road trip all you need are ballpark estimates for the price of gas, highway speeds and the distances you're traveling. Exact numbers wouldn't even be useful, since it's impossible to know the exact highway conditions anyway.

Similarly if you're drawing up a household budget you won't allocate money down to the penny. Something unforeseen always comes up, so you compensate by building in a margin of safety. If you are regularly in a position where you need to plan out your budget literally down to the penny, then exact arithmetic is probably the least of your problems.

The beauty of approximate arithmetic is that it builds up our numerical intuitions, gradually expanding our intuitive scope. You start to develop a feel for how much it costs to road-trip from Seattle to Cleveland, or the typical interest rates on a car loan, or the number of cookies you need to make for a bake sale. Physicists are famous for carrying approximate arithmetic to the extreme. They make a parlor game out of estimating how many drops of water are in Lake Michigan, the cost of sending an astronaut into space, or other such brain teasers.

I think this is the best way to teach arithmetic: keeping the material as close to our intuitions as possible. Obviously we want kids to be able to add, but we should start by focusing on the things they're naturally good at. First they should understand the material in the native language of the human brain, and only then can they learn the much less natural language of formal math. And yes, if push comes to shove and we are forced to make tradeoffs, it's more important that students can estimate a household budget than that they learn how to do long division.

This same logic applies to algebra, though it grieves me to say it. As a physics major I have poured countless hours of my life into algebra, and at this point I love the stuff (laugh if you wish). But the reality is that very few people need to master this skillset. How to factor polynomials should not be made into a stumbling block toward more generally useful knowledge, like how to interpret statistics, calculate a mortgage rate, or program a computer.

A lot of people might criticize what I'm saying as being too touchy-feely, a slippery slope toward classrooms where the value of 2+2 is negotiable. They think that math being unnatural is kinda the whole point: we need to teach students how to think rigorously!

In a way I couldn't agree more with this sentiment. I'm not trying to turn math into a subject where we all sit around and talk about our feelings. On the contrary I want to drastically increase the emphasis on algorithmic thinking, which is every bit as black-and-white as arithmetic. If your algorithm doesn't work correctly then you're as wrong as if you'd said that 2+2=7. The difference is that arithmetic teaches us to follow mindless rules, whereas algorithm design teaches us to formulate those rules by critical thinking. I want to teach students to program computers, not become computers!

Thank you for taking the time to go through this book with me. My ostensible goal has been to convince you that math isn't some arbitrary symbol game, nor is it a mystical art: it is fundamentally just a branch of language, like the English that I'm writing in. That is one key insight. The other one is that, unlike most daily speech, math tends to push the limits of our intuitive scope. If you know these two points then you can understand how mathematics evolved, what it's used for, how to teach it, and where it will take us in the future. My argument might not have fully won you over, but I hope that at least I've convinced you this question is interesting, and introduced you to some exciting new ideas.

I tried to make sure that each chapter would have something cool and new for everybody. It wasn't hard! Part of the beauty of math is that you can use it to describe just about anything: science, economics, public opinion, and even mathematics itself. This is perhaps the single biggest reason that I love math; it opens the door to innumerable interesting topics, and lets me understand them with a clarity and precision that's hard to get any other way.

If you take away only one message please make it this: in the same way that a picture is worth a thousand words, mathematics never tells us the whole story. It's easy to be lulled by the clean precision of numbers and formulas, spat out by a computer or calculated by hand. They are incredibly useful, but the idealizations behind them cover over a wealth of detail and ambiguity. As society relies increasingly on mathematical descriptions of the world, we must be ever more mindful of the crucial details that get lost in translation.

Because at the end of the day we're talking about the messy, irregular, and splendidly complex world you see around you. Is any language good enough to capture it?

NOTES ON FIGURES

The cover art was adapted from http://commons.wikimedia.org/wiki/File:Human_evolution_scheme.svg
The other images were available in the public domain (usually on Wikipedia), or I generated them myself.

RECOMMENDED READING

The Number Sense

By Stanislas DeHaene

If you want to learn more about the low-level ways that the brain processes math then this book is the place to go. Dr. DeHaene is one of the leading researchers in numerical cognition, as well as a gifted writer for lay audiences, and The Number Sense was one of my most valuable resources in writing this book.

The Language Instinct

By Steven Pinker

One of the greatest "popular intellectual" books of all time, The Language Instinct is an engaging crash course in modern linguistics, primarily from the Chomskyan school of thought. Pinker goes into a lot of detail about grammar, expanding on what I covered in chapter 2. However, he also dives into the many fascinating aspects of natural language that have no real parallel in mathematics; physiology of the vocal system, language acquisition, the surprising ubiquity of profanity, and so on. It's a great read.

Thinking, Fast and Slow

Daniel Kahneman

If you only read one popular science book from recent years, this is perhaps the best one. Kahneman is an experimental psychologist, and he divides the mind into "System 1" and "System 2". System 1 is the subconscious, instinctive part of your mind. It is constantly processing our sensory data, making lightning-fast inferences about the world, and making suggestions to our conscious mind. The number sense, for example, is mostly processed by system 1. System 2 is our conscious mind. It is deliberate and logical, but it is also painfully slow and extremely lazy, so that it almost always accepts the inferences of system 1 without further scrutiny (this is the source of our cognitive biases; system 1 is usually right, but it is also wrong in systematic ways).

Much of What is Math? can be re-cast in the language of this book. going beyond our intuitive scope corresponds to thinking about things that are beyond system 1's ability, forcing our minds to use system 2. The key to learning math correctly is not to amass facts about it, but to develop the necessary mathematical intuitions into system 1.

How Mathematics Happened; the First 50,000 Years

By Peter Rudman

I'll admit it; this book is a tad bit dry. However, it's an extremely thorough accounting of exactly how mathematics developed in early societies (as nearly as can be reconstructed). He talks about all the major number systems, the logistics of how to do arithmetic in them all, and how they probably developed. It's incredible and eye-opening to see all the ingenious tricks that went into ancient math, and how it evolved organically from basic human behaviors.

The Feynman Lectures

By Richard Feynman

My gmail status once boasted "Field Cady is now intelligent enough to understand the Feynman lectures". I used these textbooks in college but didn't really appreciate them. Years later though, I went back and realized that they are pure genius. Feynman covers everything from quantum mechanics to tornados, stripping out as much math as possible and explaining the conceptual core of the subject. It's the best example I know of showing how mathematics at its best is just critical thinking that has been translated to numerical formalism.

The books are light on math (at least, as light on math as you can make a serious physics textbook) but heavy on concepts; professional physicists read them to try and absorb some of Feynman's clarity of thought. I shouldn't be recommending these books to laymen, but they are so good I just couldn't resist. If you want to deeply understand physics, you're willing to study hard, and you want a tour through the mind of a genius, these books are for you.

The Essays of Warren Buffett

By Warren Buffett and Lawrence Cunningham

Warren Buffett, the great investor who I discussed in the book, also has a rare talent for explaining technical, usually-mathematical subjects without jargon.

This book consists of excerpts from Buffett's annual letters to Berkshire Hathaway shareholders. They're mostly concerned with business and stock investing, but there is also fantastic exposition of many key economic concepts. A bit reminiscent of Feynman, Buffett has a gift for cutting through traditional mathematical jargon and explaining the (often surprisingly simple, when you see it) core concept behind many economic ideas.

Made in the USA
San Bernardino, CA
12 August 2015